STOCHASTIC DYNAMIC MACROECONOMICS

STOCHASTIC DYNAMIC MACROECONOMICS

Theory and Empirical Evidence

Gang Gong
Willi Semmler

OXFORD
UNIVERSITY PRESS

2006

OXFORD
UNIVERSITY PRESS

Oxford University Press, Inc., publishes works that further
Oxford University's objective of excellence
in research, scholarship, and education.

Oxford New York
Auckland Cape Town Dar es Salaam Hong Kong Karachi
Kuala Lumpur Madrid Melbourne Mexico City Nairobi
New Delhi Shanghai Taipei Toronto

With offices in
Argentina Austria Brazil Chile Czech Republic France Greece
Guatemala Hungary Italy Japan Poland Portugal Singapore
South Korea Switzerland Thailand Turkey Ukraine Vietnam

Published by Oxford University Press, Inc.
198 Madison Avenue, New York, New York 10016

www.oup.com

Library of Congress Cataloging-in-Publication Data
Gong, Gang, 1959–
Stochastic dynamic macroeconomics : theory and empirical evidence /
Gang Gong and Willi Semmler.
p. cm.
Includes bibliographical references and index.
ISBN-13 978-0-19-530162-5
ISBN 0-19-530162-5
1. Macroeconomics—Mathematical models. 2. Equilibrium (Economics)—Mathematical models. 3. Business cycles—Mathematical models. 4. Keynesian economics—Mathematical models. 5. Stochastic analysis. I. Semmler, Willi. II. Title.

HB172.5.G66 2005
339'.01'51922—dc22 2005049870

9 8 7 6 5 4 3 2 1

Printed in the United States of America
on acid-free paper

Our present is determined as much by where we want to go as it is by where we have come from.

—Ernst Bloch

Preface

This book aims to contribute to the study of alternative paradigms in macroeconomics. As with other recent approaches to dynamic macroeconomics, we build on intertemporal economic behavior of economic agents, but stress Keynesian features more than other recent literature in this area. In general, stochastic dynamic macromodels are difficult to solve and to estimate, particularly if intertemporal behavior of economic agents is involved. Thus, besides addressing important macroeconomic issues in a dynamic framework, a major focus of this book is to discuss and apply solutions and estimation methods to models with intertemporal behavior of economic agents.

The material of this book has been presented by the authors at several universities. Chapters of the book have been presented as lectures at Bielefeld University; Foscari University, Venice; the University of Technology, Vienna; the University of Aix-en-Provence; Columbia University, New York; New School University, New York; Bejing University; Tsinghua University, Bejing; the Chinese University of Hong Kong; the City University of Hong Kong; the European Central Bank; and the Deutsche Bundesbank. Some chapters of the book also have been presented at the annual conferences of the American Economic Association, the Society of Computational Economics, and the Society of Nonlinear Dynamics and Econometrics.

We are grateful for comments by the participants at those conferences. We are grateful for discussions with Philippe Aghion, Toichiro Asada, Jean-Paul Bénassy, Peter Flaschel, Buz Brock, Lars Grüne, Richard Day, Ekkehard Ernst, Ray Fair, Stefan Mittnik, James Ramsey, Jens Rubart, Volker Caspari,

Malte Sieveking, Michael Woodford, and colleagues at our universities. We thank Uwe Köller for research assistance, and Gaby Windhorst for editing and typing the manuscript. Financial support from the Ministry of Education, Science, and Technology of the State of North Rhine-Westphalia, Germany, and from the Bernhard Schwartz Center of Economic Policy Analysis of the New School University is gratefully acknowledged. Finally, we want to thank anonymous referees and Terry Vaughn of Oxford University Press, who have helped the book to become better.

Contents

STOCHASTIC DYNAMIC MACROECONOMICS

Introduction and Overview

The dynamic general equilibrium (DGE) model, in particular its more popular version, the real business cycle (RBC) model, has become a major paradigm in macroeconomics. It has been applied in numerous fields of economics. Its essential features are the assumptions of intertemporal optimizing behavior of economic agents, competitive markets, and price-mediated market clearing through flexible wages and prices. In this type of stochastic dynamic macromodeling, only real shocks, such as technology shocks, monetary and government spending shocks, variations in tax rates, or shifts in preferences generate macro fluctuations.

Recently Keynesian features have been built into the DGE model by preserving its characteristics, such as intertemporally optimizing agents and market clearing, but introducing monopolistic competition and sticky prices and wages. In particular, in numerous papers and in a recent book Woodford (2003) has worked out this new paradigm in macroeconomics, which is now commonly called New Keynesian macroeconomics. In contrast to the traditional Keynesian macromodels, such variants also presume dynamically optimizing agents and market clearing, but sluggish wage and price adjustments.[1]

The standard DGE model fails to replicate essential product, labor market, and asset market characteristics. In our book—different from the DGE model and its competitive or monopolistic variants—we do not presume the clearing of all markets in all periods. As in the monopolistic competition variant of the DGE model, we permit nominal rigidities. Yet, by stressing

Keynesian features in a model with production and capital accumulation, we demonstrate that even with dynamically optimizing agents, not all markets may be cleared.

Solution and Estimation Methods

Whereas models with Keynesian features are worked out and stressed in part III, parts I and II provide the groundwork for them. In parts I and II we build extensively on the basics of stochastic dynamic macroeconomics.

Part I is the technical preparation for the development of our theoretical arguments. Here we provide a variety of technical tools to solve and estimate stochastic dynamic optimization models, which is a prerequisite for a proper empirical assessment of the models we treat. Solution methods are presented in chapters 1 and 2, and estimation methods, along with calibration and the current methods of empirical assessment, are introduced in chapter 3. These methods are applied in the remaining chapters of the book.

Solving stochastic dynamic optimization models has been an important research topic since the mid-1990s, and many different methods have been proposed. Usually an exact and analytical solution of a dynamic decision problem is not attainable. Therefore one has to rely on an approximate solution, which may also have to be computed by numerical methods. Recently, numerous methods have been developed to solve stochastic dynamic decision problems. Among the well-known methods are the perturbation and projection methods (Judd 1998), the parameterized expectations approach (den Haan and Marcet 1990), and the dynamic programming approach (Santos and Vigo-Aguiar 1998; Grüne and Semmler 2004a). When an exact and analytical solution to a dynamic optimization problem is not attainable, one has to use numerical methods. A solution method with higher accuracy often requires more complicated procedures and longer computation time.

In this book, in order to allow for an empirical assessment of stochastic dynamic models, we focus on approximate solutions that are computed from two types of first-order conditions: the Euler equation and the equation derived from the Lagrangian. For these two types of first-order conditions, three types of approximation methods can be found in the literature: the Fair–Taylor method, the log-linear approximation method, and the linear-quadratic approximation method. After a discussion of the various approximation methods, we introduce a method that will be used repeatedly in the subsequent chapters. The method, which has been written into a "GAUSS" procedure, has the advantage of short computation time and easy implementation without sacrificing too much accuracy. We will also compare those approximation methods with the dynamic programming approach.

Often the methods use a smooth approximation of first-order conditions, such as the Euler equation. Sometimes, smooth approximations are not useful if the value function is very steep or if it is not differentiable and thus non-smooth. A method such as that employed by Grüne and Semmler (2004a) can then be used.

Less progress has been made regarding the empirical assessment and estimation of stochastic dynamic models. Given the wide application of stochastic dynamic models expected in the future, we believe that the estimation of such models will become an important research topic. The discussion in chapters 3–6 can be regarded as an important step toward that purpose. Our proposed estimation strategy requires solving the stochastic dynamic optimization model repeatedly, at various possible structural parameters searched by a numerical algorithm within the parameter space. This requires that the solution methods adopted in the estimation strategy should consume as little time as possible while not losing too much accuracy. After comparing different approximation methods, we find the proposed methods of solving stochastic dynamic optimization models, such as those used in chapters 3–6, most useful. We also will explore the impact of the use of different data sets on the calibration and estimation results.

RBC Model as a Benchmark

In part II, we set up a benchmark model, the RBC model, for comparison in terms of either theory or empirics.

The standard RBC model is a representative agent model, but it is constructed on the basis of neoclassical general equilibrium theory. It therefore assumes that all markets (including product, capital, and labor markets) are cleared in all periods, regardless of whether the model refers to the short run or the long run. The imposition of market clearing requires that prices be set at an equilibrium level. At the pure theoretical level, the existence of such general equilibrium prices can be proved under certain assumption. Little, however, has been said about how the general equilibrium can be achieved. In an economy in which both firms and households are price takers, an auctioneer who adjusts the price toward some equilibrium is implicitly presumed to exist. Thus, how an equilibrium is brought about is essentially a Walrasian *tâtonnement* process.

Working with such a framework of competitive general equilibrium is elegant and perhaps a convenient starting point for economic analysis. It nevertheless neglects many restrictions on the behavior of agents: the trading process and the market-clearing process, the implementation of new technology, and the market structure, among many others. In part II, we provide a thorough review of the standard RBC model, the representative stochastic dynamic model of the competitive general equilibrium type. The review starts with laying out a microfoundation, and then discusses a variety of empirical issues, such as the estimation of structural parameters, the data construction, the matching with the empirical data, the asset market implications, and so on. The issues explored in this part of the book provide the incentives to introduce Keynesian features into a stochastic dynamic model as developed in part III. It also provides a reasonable ground on which to judge new model variants by considering whether they can resolve some puzzles, as explored in part II.

Open-Ended Dynamics

One of the restrictions in the standard RBC model is that the firm does not face any additional cost (a cost beyond the usual activities at the current market prices) when it adjusts either price or quantity. For example, changing the price may require the firm to pay a menu cost and also, more important, a reputation cost. It is the cost arising from price and wage adjustments that has become an important focus of New Keynesian research over the last decades.[2] However, there may also be an adjustment cost on the real side arising from a change in quantity. In a production economy, increasing output requires the firm to hire new workers and add new capacity. In a given period of time, a firm may find it increasingly difficult to create additional capacity. This indicates that there will be an adjustment cost in creating capacity (or capital stock, via investment) and, further, that such adjustment cost may also be an increasing function of the size of investment.

In chapter 7, we introduce adjustment costs into the benchmark RBC model. This may bring about multiple equilibria toward which the economy may move. The dynamics are open ended in the sense that the economy can move to a low level or a high level of activity.[3] Such an open-ended dynamics is certainly one of the important feature of Keynesian economics. In recent times such open-ended dynamics have been found in a large number of dynamic models with intertemporal optimization. Those models have been called indeterminacy and multiple equilibria models. Theoretical models of this type are studied in Benhabib and Farmer (1999) and Farmer (1999), and an empirical assessment is given in Schmidt-Grohe (2001). Some of the models are RBC models with increasing returns to scale and/or more general preferences than power utility that generates indeterminacy. Local indeterminacy and a global multiplicity of equilibria can arise here. Other models are monetary macromodels, where consumers' welfare is affected positively by consumption and cash balances, and negatively by the labor effort and an inflation gap from some target rates. For certain substitution properties between consumption and cash holdings, those models admit unstable as well as stable high-level and low-level steady states. There also can be indeterminacy in the sense that any initial condition in the neighborhood of one of the steady states is associated with a path toward, or away from, that steady state (see Benhabib et al. 2001).

Overall, the indeterminacy and multiple equilibria models predict an open-ended dynamics, arising from sunspots, in which the sunspot dynamics are frequently modeled by versions with multiple steady-state equilibria, and there are also pure attractors (repellers), permitting any path in the vicinity of the steady-state equilibria to move back to (away from) the steady-state equilibrium. Although these are important variants of macrodynamic models with optimizing behavior, it has recently been shown that indeterminacy is likely to occur within only a small set of initial conditions.[4] Yet, despite such unsolved problems, the literature on open-ended dynamics has greatly enriched macrodynamic modeling.

Pursuing this line of research, we introduce in chapter 7 a simple model in which one does not need to refer to model variants with externalities (and increasing returns to scale) and/or to more elaborate preferences in order to obtain such results. We show that through the adjustment cost of capital we may obtain non-uniqueness of steady-state equilibria in an otherwise standard dynamic optimization version. Multiple steady-state equilibria, in turn, lead to thresholds separating different domains of attraction of capital stock, consumption, employment, and welfare level. As our solution shows, thresholds are important as separation points below or above which it is advantageous to move to lower or higher levels of capital stock, consumption, employment, and welfare. Our model version thus can explain how the economy becomes history dependent and moves, after a shock or policy influences, to low- or high-level equilibria in employment and output.

Nonclearing Markets

A second important feature of Keynesian macroeconomics concerns the modeling of the labor market. An important characteristic of the DGE model is that it is a market-clearing model. For the labor market, the DGE model predicts an excessive smoothness of labor effort in contrast to empirical data. The low variation in the employment series is a well-known puzzle in the RBC literature.[5] It is related to the specification of the labor market as a cleared market. Although in its structural setting (see, e.g., Stokey et al. 1989) the DGE model specifies both sides of a market, demand and supply, the moments of the macro variables of the economy are generated by a one-sided force due to its assumption on wage and price flexibility; and, thus, on equilibrium in all markets, including output, labor, and capital markets. The labor effort results only from the decision rule of the representative agent to supply labor. In our view, there should be no restriction for the other side of the market, the demand, to have effects on the variation of labor effort.

Attempts have been made to introduce imperfect competition features into the DGE model.[6] In those types of models, producers set the price optimally, according to their expected market demand curve. If one follows a Calvo price-setting scheme, there will be a gap between the optimal price and the existing price. However, it is presumed that the market is still cleared, since the producer is assumed to supply the output according to what the market demands at the existing price. This consideration also holds for the labor market. Here the wage rate is set optimally by the household, according to the expected market demand curve for labor. Once the wage has been set, it is assumed to be rigid (or adjusted slowly). Thus, if the expectation is not fulfilled, there will be a gap between the optimal wage and the existing wage. Yet in the New Keynesian models the market is still assumed to be cleared, since the household is assumed to supply labor regardless of the demand at the given wage rate. (Yet, as we will discuss further in chapter 8, this definition of market clearing is not unambiguous.)

In order to better fit the RBC model's predictions to the labor market data, search and matching theory has been employed to model the labor market in the context of an RBC model. (For further details, see chapter 8.) Informational or institutional search frictions may then explain equilibrium unemployment rates and their rise. Yet, those models still have a hard time explaining the large variation of vacancies and unemployment and the strong shift of unemployment rates, such as those experienced in Europe since the 1980s, as equilibrium unemployment rates.[7]

Concerning the labor market in Keynesian terms, we pursue an approach that allows for a nonclearing labor market. In our view, the decisions with regard to price and quantities can be made separately, both subject to optimal behavior. When the price has been set, and is sticky for a certain period, the price is given to the supplier deciding on the quantities. There is no reason why the firm cannot choose the optimal quantity rather than what the market demands, especially when the optimum quantity is less than the quantity demanded by the market. This consideration will allow for nonclearing markets.[8] Our proposed new model helps to study labor market problems by being based on adaptive optimization in which households, after a first round of optimization, have to reoptimize when facing constraints on supplying labor to the market. On the other hand, firms may face constraints on the product markets. As we will show in chapters 8 and 9, such a multiple-stage optimization model allows for greater volatility of the employment rates compared with the standard RBC model, and also provides, a framework to study the secular rise or fall of unemployment.

Technology and Demand Shocks

A further Keynesian feature of macromodels concerns the role of shocks. In the standard DGE model, technology shocks are the driving force of the business cycles and are assumed to be measured by the Solow residual. Since the Solow residual is computed on the basis of observed output, capital, and employment, it is presumed that all factors are fully utilized. However, there are several reasons to distrust the standard Solow residual as a measure of technology shock. First, Mankiw (1989) and Summers (1986) have argued that such a measure often leads to excessive volatility in productivity, and even the possibility of technological regress, both of which seem to be empirically implausible. Second, it has been shown that the Solow residual can be expressed by some exogenous variables—for example, demand shocks arising from military spending (Hall 1988) and changed monetary aggregates (Evans 1992)—that are unlikely to be related to factor productivity. Third, the standard Solow residual can be contaminated if the cyclical variations in factor utilization are significant.

Since the Solow residual cannot be trusted as a measure of technology shock, researchers have developed different methods to measure technology shocks correctly. All these methods are focused on the computation of factor utilization. There are basically three strategies. The first is to use an observed indicator as proxy for unobserved utilization. A typical example is to employ electricity use as a proxy for capacity utilization (see Burnside et al. 1996). Another strategy is to construct an economic model so that one can compute the factor utilization from the observed variables (see Basu and Kimball 1997; Basu et al. 1998). A third strategy uses an appropriate restriction in a VAR estimate to identify a technology shock (see Gali 1999; Francis and Ramey 2001, 2003).

One of the major celebrated arguments of RBC theory is that technology shocks are pro-cyclical. A positive technology shock will increase output, consumption, and employment. Yet this result is obtained from the empirical evidence, in which the technology shock is measured by the standard Solow residual. Like Gali (1999) and Francis and Ramey (2001, 2003), we find that if one uses the corrected Solow residual, the technology shock is negatively correlated with employment, and therefore the RBC model loses its major driving force (see chapters 5 and 9).

Puzzles to Be Solved

To sum up, we may say that the standard RBC model has left us with major puzzles. The first type of puzzle is related to the asset market and is often discussed under the heading of the equity premium puzzle. Extensive research has attempted to improve on this problem by elaborating on more general preferences and technology shocks. Chapter 6 studies in detail the asset price implication of the RBC model. The second puzzle is, as mentioned above, related to the labor market. The RBC model generally predicts an excessive smoothness of labor effort and wrong cross-correlation of labor effort with other macrovariables, in contrast to empirical data. Third, the RBC model predicts a significantly high positive correlation between technology and employment, whereas empirical research demonstrates, at least for business cycle frequency, a negative or almost 0 correlation. This has been named the technology puzzle. The first puzzle is studied in chapter 6 of this volume, and chapters 8 and 9 are mainly concerned with the latter two puzzles.

Of course numerous other puzzles arise when such an intertemporal decision framework is applied to macroeconomics. Particular problems arise—for example, for modeling asset markets and monetary and fiscal policy, or open economy issues that we do not discuss in this volume. Finally, future work on intertemporal decision models will face new challenges arising from the problem of the accuracy of the solution methods used when moving toward empirical applications of large-scale models.

PART I

SOLUTION AND ESTIMATION OF STOCHASTIC DYNAMIC MODELS

1

Solution Methods of Stochastic Dynamic Models

1.1 Introduction

The dynamic decision problem of an economic agent whose objective is to maximize his or her utility over an infinite time horizon is often studied in the context of a stochastic dynamic optimization model. To understand the structure of this decision problem, we describe it in terms of a recursive decision problem of a dynamic programming approach. Thereafter, we discuss some solution methods frequently employed to solve the dynamic decision problem.

In most cases, an exact and analytical solution of the dynamic programming decision is not attainable. Therefore one has to rely on an approximate solution, which may also have to be computed by numerical methods. Recently, numerous methods to solve stochastic dynamic decision problems have been developed. Among the well-known methods are the perturbation and projection methods (Judd 1996), the parameterized expectations approach (den Haan and Marcet 1990) and the dynamic programming approach (Santos and Vigo-Aguiar 1998; Grüne and Semmler 2004a). In this book, in order to allow for an empirical assessment of stochastic dynamic models, we focus on approximate solutions that are computed from two types of first-order conditions: the Euler equation and the equation derived from the Lagrangian. Given these two types of first-order conditions, three types of approximation methods can be found in the literature: the Fair–Taylor method, the log-linear approximation method, and the linear-quadratic approximation method. Still, in most of the cases, an approximate solution cannot be derived analytically, and therefore a numerical algorithm is needed to facilitate the computation of the solution.

In this chapter, we discuss those various approximation methods and then propose another numerical algorithm that can help us to compute approximate solutions. The algorithm is used to compute the solution path obtained from the method of linear-quadratic approximation with the first-order condition derived from the Lagrangian. While the algorithm takes full advantage of an existing procedure, it overcomes the limitations given by the Chow (1993a, 1993b) method.

The remainder of this chapter is organized as follows. In section 1.2 we present the standard recursive method, which uses the value function for iteration. We show that the standard recursive method may encounter difficulties when applied to compute a dynamic model. Section 1.3 establishes the two first-order conditions to which different approximation methods can be applied. Section 1.4 briefly reviews the various approximation methods in the existing literature. Section 1.5 presents our new algorithm for dynamic optimization. Appendix I provides the proof of the propositions in the text. Finally, a GAUSS procedure that implements our suggested algorithm is presented in Appendix II.

1.2 The Standard Recursive Method

We consider a representative agent whose objective is to find a control (or decision) sequence $\{u_t\}_{t=0}^{\infty}$ such that

$$\max_{\{u\}_{t=0}^{\infty}} E_0 \left[\sum_{t=0}^{\infty} \beta^t U(x_t, u_t) \right], \tag{1.1}$$

subject to

$$x_{t+1} = F(x_t, u_t, z_t). \tag{1.2}$$

Above, x_t is a vector of m state variables at period t; u_t is a vector of n control variables; z_t is a vector of s exogenous variables whose dynamics does not depend on x_t and u_t; E_t is the mathematical expectation conditional on the information available at time t; and $\beta \in (0, 1)$ denotes the discount factor.

Let us first make several remarks regarding the formulation of the above problem. First, this formulation assumes that the uncertainty of the model comes only from the exogenous z_t. One popular assumption regarding the dynamics of z_t is to assume that z_t follow an $AR(1)$ process:

$$z_{t+1} = Pz_t + p + \epsilon_{t+1}, \tag{1.3}$$

where ϵ_t is independently and identically distributed (i.i.d.). Second, this formulation is not restricted to those structure models with more lags or leads. A model with finite lags or leads can be transformed through the use of auxiliary variables into an equivalent model with one lag or lead. Third, the initial condition (x_0, z_0) in this formulation is assumed to be given.

The problem with solving a dynamic decision problem is to seek a time-invariant policy function G mapping from the state and exogenous (x, z) into the control u. With such a policy function (or control equation), the sequences of state $\{x_t\}_{t=1}^{\infty}$ and control $\{u_t\}_{t=0}^{\infty}$ can be generated by iterating the control equation

$$u_t = G(x_t, z_t), \tag{1.4}$$

as well as the state equation (1.2), given the initial condition (x_0, z_0) and the exogenous sequence $\{z_t\}_{t=1}^{\infty}$ generated by (1.3).

To find the policy function G by the recursive method, we first define a value function V:

$$V(x_0, z_0) = \max_{\{u_t\}_{t=0}^{\infty}} E_0 \left[\sum_{t=0}^{\infty} \beta^t U(x_t, u_t) \right]. \tag{1.5}$$

Expression (1.5) can be transformed to reveal its recursive structure. For this purpose, we first rewrite (1.5) as follows:

$$\begin{aligned} V(x_0, z_0) &= \max_{\{u_t\}_{t=0}^{\infty}} \left\{ U(x_0, u_0) + E_0 \left[\sum_{t=1}^{\infty} \beta^t U(x_t, u_t) \right] \right\} \\ &= \max_{\{u_t\}_{t=0}^{\infty}} \left\{ U(x_0, u_0) + \beta E_0 \left[\sum_{t=0}^{\infty} \beta^t U(x_{t+1}, u_{t+1}) \right] \right\}. \end{aligned} \tag{1.6}$$

It is easy to find that the second term in (1.6) can be expressed as β times the value V as defined in (1.5) with the initial condition (x_1, z_1). Therefore, we can rewrite (1.5) as

$$V(x_0, z_0) = \max_{\{u_t\}_{t=0}^{\infty}} \{U(x_0, u_0) + \beta E_0 [V(x_1, z_1)]\}. \tag{1.7}$$

The formulation of equation (1.7) represents a dynamic programming problem that highlights the recursive structure of the decision problem. In every period t, the planner faces the same decision problem: choosing the control variable u_t that maximizes the current return plus the discounted value of the optimum plan from period $t + 1$ onward. Since the problem repeats itself every period, the time subscripts become irrelevant. We thus can write (1.7) as

$$V(x,z) = \max_u \{U(x,u) + \beta E[V(\tilde{x},\tilde{z})]\}, \tag{1.8}$$

where the tildes (∼) over x and z denote the corresponding next period values. Obviously, they are subject to (1.2) and (1.3). Equation (1.8) is also called the Bellman equation, named after Richard Bellman (1957). If we know the function V, we can solve u via the Bellman equation.

Unfortunately, all these considerations are based on the assumption that we know the function V, which in reality we do not know in advance. The typical method in this case is to construct a sequence of value functions by iterating the following equation:

$$V_{j+1}(x,z) = \max_u \left\{U(x,u) + \beta E\left[V_j(\tilde{x},\tilde{z})\right]\right\}. \tag{1.9}$$

In terms of an algorithm, the method can be described as follows:

- Step 1. Guess a differentiable and concave candidate value function, V_j.
- Step 2. Use the Bellman equation to find the optimum u and then compute V_{j+1} according to (1.9).
- Step 3. If $V_{j+1} = V_j$, stop. Otherwise, update V_j and go back to step 1.

Under some regularity conditions regarding the functions U and F, the convergence of this algorithm is warranted by the contraction-mapping theorem (Sargent 1987; Stokey et al. 1989). However, the difficulty of this algorithm is that in each step 2, we need to find the optimum u that maximizes the right side of equation (1.9). This task makes it difficult to write a closed-form algorithm for iterating the Bellman equation. Researchers are therefore forced to seek different numerical approximation methods.

1.3 The First-Order Conditions

Since the mid-1980s there have been various methods of numerical approximation used to solve the problem of dynamic optimization.[1] As stated above, in this book we want to focus on the first-order conditions that are used to derive the decision sequence. There are two approaches: one is to use the Euler equation, and the other, the equation derived from the Lagrangian.

1.3.1 The Euler Equation

We start from the Bellman equation (1.8). The first-order condition for maximizing the right side of the equation takes the form

$$\frac{\partial U(x,u)}{\partial u} + \beta E\left[\frac{\partial F}{\partial u}(x,u,z)\frac{\partial V}{\partial \tilde{x}}(\tilde{x},\tilde{z})\right] = 0. \tag{1.10}$$

The objective here is to find $\partial V/\partial x$. Assume V is differentiable, and thus, from (1.8), it satisfies

$$\frac{\partial V(x,z)}{\partial x} = \frac{\partial U(x, G(x,z))}{\partial x} + \beta E\left[\frac{\partial F}{\partial x}(x, G(x,z), z)\frac{\partial V}{\partial F}(\tilde{x}, \tilde{z})\right]. \tag{1.11}$$

This equation is often called the Benveniste–Scheinkman formula.[2] Assume $\partial F/\partial x = 0$. The above formula then becomes

$$\frac{\partial V(x,z)}{\partial x} = \frac{\partial U(x, G(x,z))}{\partial x}. \tag{1.12}$$

Substituting this formula into (1.10) gives rise to the Euler equation:

$$\frac{\partial U(x,u)}{\partial u} + \beta E\left[\frac{\partial F}{\partial u}(x, u, z)\frac{\partial U}{\partial \tilde{x}}(\tilde{x}, \tilde{u})\right] = 0, \tag{1.13}$$

where the tilde ($\sim$) over u denotes the next period value with respect to u.

Note that in order to use the Euler equation as the first-order condition for deriving the decision sequence, $\partial F/\partial x = 0$ is required. In economic analysis, one often encounters models, after some transformation, in which x does not appear in the transition law so that $\partial F/\partial x = 0$ is satisfied. We will show this technique in chapter 2, using a prototype model as a practical example. However, there are still models in which such transformation is not feasible.

1.3.2 Deriving the First-Order Condition from the Lagrangian

Suppose that for the dynamic optimization problem as represented by (1.1) and (1.2), we can define the Lagrangian L:

$$L = E_0 \sum_{t=0}^{\infty} \left\{\beta^t U(x_t, u_t) - \beta^{t+1}\lambda'_{t+1}\left[x_{t+1} - F(x_t, u_t, z_t)\right]\right\},$$

where λ_t, the Lagrangian multiplier, is an $m \times 1$ vector. Setting the partial derivatives of L to 0 with respect to λ_t, x_t, and u_t will yield equation (1.2) as well as

$$\frac{\partial U(x_t, u_t)}{\partial x_t} + \beta E_t \lambda_{t+1} \frac{\partial F(x_t, u_t, z_t)}{\partial x_{t+1}} = \lambda_t, \tag{1.14}$$

$$\frac{\partial U(x_t, u_t)}{\partial u_t} + \beta E_t \lambda_{t+1} \frac{\partial F(x_t, u_t, z_t)}{\partial u_t} = 0. \tag{1.15}$$

In comparison with the Euler equation, we find that there is an unobservable variable λ_t appearing in the system. Yet, by using (1.14) and (1.15), one does not have to transform the model into the setting that $\partial F/\partial x = 0$.

This is an important advantage over the Euler equation. Also, as we will see in chapter 2, these two types of first-order conditions are equivalent when we appropriately define λ_t in terms of x_t, u_t, and z_t.[3] This further implies that they can produce the same steady states when being evaluated in their certainty equivalence forms.

1.4 Approximation and Solution Algorithms

1.4.1 The Gauss–Seidel Procedure and the Fair–Taylor Method

The state (1.2), the exogenous (1.3), and the first-order condition derived either as the Euler equation (1.13) or from the Lagrangian (1.14)–(1.15) form a dynamic system from which the transition sequences $\{x_{t+1}\}_{t=0}^{\infty}$, $\{z_{t+1}\}_{t=0}^{\infty}$, $\{u_t\}_{t=0}^{\infty}$, and $\{\lambda_t\}_{t=0}^{\infty}$ are implied, given the initial condition (x_0, z_0). Yet, such a system is mostly highly nonlinear, and therefore the solution paths usually cannot be computed directly. One popular approach, suggested by Fair and Taylor (1983), is to use a numerical algorithm, called the Gauss–Seidel procedure. For the convenience of presentation, the following discussion assumes that only the Euler equation is used.

Suppose that the system can be be written as the following $m+n$ equations:

$$f_1(y_t, y_{t+1}, z_t, \psi) = 0 \tag{1.16}$$

$$f_2(y_t, y_{t+1}, z_t, \psi) = 0 \tag{1.17}$$

$$\vdots$$

$$f_{m+n}(y_t, y_{t+1}, z_t, \psi) = 0. \tag{1.18}$$

Here y_t is the vector of endogenous variables with $m + n$ dimensions, including both state x_t and control u_t;[4] ψ is the vector of structural parameter. Also, in this formulation we left aside the expectation operator E. This can be done by setting the corresponding disturbance term, if there is any, to the expectation values (usually 0). Therefore the system is essentially not different from the dynamic rational expectation model considered by Fair and Taylor (1983).[5] The system, as suggested, can be solved numerically by an iterative technique, called the Gauss–Seidel procedure, to which we now turn.

It is always possible to tranform the system (1.16)–(1.18) as follows:

$$y_{1,t+1} = g_1(y_t, y_{t+1}, z_t, \psi) \tag{1.19}$$

$$y_{2,t+1} = g_2(y_t, y_{t+1}, z_t, \psi) \tag{1.20}$$

$$\vdots$$

$$y_{m+n,t+1} = g_{m+n}(y_t, y_{t+1}, z_t, \psi), \tag{1.21}$$

where, $y_{i,t+1}$ is the ith element in the vector y_{t+1}, $i = 1, 2, \ldots, m+n$. Given the initial condition $y_0 = y_0^*$, and the sequence of exogenous variables $\{z_t\}_{t=0}^T$, with T the prescribed time horizon of our problem, the algorithm starts by setting $t = 0$ and proceeds as follows:

- Step 1. Set an initial guess on y_{t+1}. Call this guess $y_{t+1}^{(0)}$. Compute y_{t+1} according to (1.19)–(1.21) for the given $y_{t+1}^{(0)}$ along with y_t. Denote this new computed value $y_{t+1}^{(1)}$.
- Step 2. If the distance between $y_{t+1}^{(1)}$ and $y_{t+1}^{(0)}$ is less than a prescribed tolerance level, go to step 3. Otherwise, compute $y_{t+1}^{(2)}$ for the given $y_{t+1}^{(1)}$. This procedure is repeated until the tolerance level is satisfied.
- Step 3. Update t by setting $t = t+1$ and go to step 1.

The algorithm continues until t reaches T. This will produce a sequence of endogenous variables $\{y_t\}_{t=0}^T$ that include both decision $\{u_t\}_{t=0}^T$ and state $\{x_t\}_{t=0}^T$ variables.

There is no guarantee that convergence can always be achieved for the iteration in each period. If this is the case, a damping technique can usually be employed to force convergence (see Fair 1984, chapter 7). The second disadvantage of this method is the cost of computation. The procedure requires the iteration and convergence for each period t, $t = 1, 2, 3, \ldots, T$. This cost of computation makes it a difficult candidate for solving the dynamic optimization problem. The third and most important problem regards its accuracy of solution. The procedure starts with the given initial condition y_0, and therefore the solution sequences $\{u_t\}_{t=1}^T$ and $\{x_t\}_{t=1}^T$ depend on the initial condition, which includes not only the initial state x_0 but also the initial decisions u_0. Yet the initial condition for the dynamic decision problem is usually provided only by x_0 (see equations (1.1–1.2) in section 1.2). Considering that the weights of u_0 could be important in the value of the objective function (1.1), there might be a problem in accuracy. One possible way to deal with this problem is to start with different initial u_0. In chapter 2, when we turn to a practical problem, we will investigate these issues more thoroughly.

1.4.2 The Log-Linear Approximation Method

Solving the nonlinear dynamic optimization model with log-linear approximation has been widely used and well documented. It has been proposed in particular by King et al. (1988a, 1988b) and Campbell (1994) in the context of RBC models. In principle, this approximation method can be applied to the first-order condition either in terms of the Euler equation or derived from the Langrangian. Formally, let X_t be the variables and $\bar{X}$ the corresponding steady state. Then,

$$x_t \equiv \ln X_t - \ln \bar{X}$$

is regarded to be the log-deviation of X_t. In particular, $100x_t$ is the percentage by which X_t deviates from $\bar{X}$. The general idea of this method is to replace all the necessary equations in the model by approximations, which are linear in the log-deviation form. Given the approximate log-linear system, one then uses the method of undetermined coefficients to solve for the decision rule, which is also in the form of log-linear deviations.

Uhlig (1999) provides a tool kit of such a method for solving a general dynamic optimization model. The general procedure involves the following steps:

- Step 1. Find the necessary equations characterizing the equilibrium law of motion of the system. These equations should include the state equation (1.2), the exogenous equation (1.3), and the first-order condition derived either as the Euler equation (1.13) or from the Lagrangian (1.14) and (1.15).
- Step 2. Derive the steady state of the model. This requires first parameterizing the model and then evaluating the model at its certainty equivalence form.
- Step 3. Log-linearize the necessary equations characterizing the equilibrium law of motion of the system. Uhlig (1999) suggests the following building blocks for such log-linearization.

$$X_t \approx \bar{X}e^{x_t} \tag{1.22}$$

$$e^{x_t + ay_t} \approx 1 + x_t + ay_t \tag{1.23}$$

$$x_t y_t \approx 0. \tag{1.24}$$

- Step 4. Solve the log-linearized system for the decision rule (which is also in log-linear form) with the method of undetermined coefficients.

In chapter 2, we will provide a concrete example to which to apply the above procedure and solve a practical problem of dynamic optimization.

Solving a nonlinear dynamic optimization model with log-linear approximation usually does not require extensive computation, in contrast to the Fair–Taylor method. In some cases, the decision rule can be derived analytically. On the other hand, by assuming a log-linearized decision rule as expressed in (1.4), the solution path does not require the initial condition of u_0, and therefore it should be more accurate than the Fair–Taylor method. However, log-linearization and solving for the undetermined coefficients are not easy, and usually have to be accomplished by hand. It is desirable to have a numerical algorithm available that can take over at least part of the analytical derivation process.

1.4.3 Linear-Quadratic Approximation with Chow's Algorithm

Another important approximation method is the linear-quadratic approximation. Again, in principle this method can be applied to the first-order condition either in terms of the Euler equation or derived from the Lagrangian. Chow (1993) was among the first to solve a dynamic optimization model with linear-quadratic approximation applied to the first-order condition derived from the Lagrangian. At the same time, he proposed a numerical algorithm to facilitate the computation of the solution.

Chow's method can be presented in both continuous and discrete time forms. Since the models in discrete time are more convenient for empirical and econometric studies, here we consider only the discrete time version. The numerical properties of this approximation method have further been studied in Reiter (1997) and Chow and Kwan (1997).

Suppose the objective of a representative agent can be written as (1.1), but subject to

$$x_{t+1} = F(x_t, u_t) + \epsilon_{t+1}. \tag{1.25}$$

The state equation here is slightly different from the one expressed by equation (1.2). Apparently, it is only a special case of (1.2). Consequently, the Lagrangian L should be defined as

$$L = E_0 \sum_{t=0}^{\infty} \left\{ \beta^t U(x_t, u_t) - \beta^{t+1} \lambda'_{t+1} \left[x_{t+1} - F(x_t, u_t) - \epsilon_{t+1} \right] \right\}.$$

Setting the partial derivatives of L to 0 with respect to λ_t, x_t, and u_t will yield equation (1.25) as well as

$$\frac{\partial U(x_t, u_t)}{\partial x_t} + \beta \frac{\partial F(x_t, u_t)}{\partial x_t} E_t \lambda_{t+1} = \lambda_t, \tag{1.26}$$

$$\frac{\partial U(x_t, u_t)}{\partial u_t} + \beta \frac{\partial F(x_t, u_t)}{\partial u_t} E_t \lambda_{t+1} = 0. \tag{1.27}$$

The linear-quadratic approximation assumes that the state equations are linear and that the objective function is quadratic. In other words,

$$\frac{\partial U(x_t, u_t)}{\partial x_t} = K_1 x_t + K_{12} u_t + k_1 \tag{1.28}$$

$$\frac{\partial U(x_t, u_t)}{\partial u_t} = K_2 u_t + K_{21} x_t + k_2 \tag{1.29}$$

$$F(x_t, u_t) = A x_t + C u_t + b + \epsilon_{t+1}. \tag{1.30}$$

Given this linear-quadratic assumption, equations (1.26) and (1.27) can be rewritten as

$$K_1 x_t + K_{12} u_t + k_1 + \beta A' E_t \lambda_{t+1} = \lambda_t \tag{1.31}$$

$$K_2 u_t + K_{21} x_t + k_2 + \beta C' E_t \lambda_{t+1} = 0. \tag{1.32}$$

Assume that the transition laws of u_t and λ_t take the linear forms

$$u_t = G x_t + g \tag{1.33}$$

$$\lambda_{t+1} = H x_{t+1} + h. \tag{1.34}$$

Chow (1993b) proves that the coefficient matrices G and H and the vectors g and h satisfy

$$G = -\left(K_2 + \beta C' H C\right)^{-1} \left(K_{21} + \beta C' H A\right) \tag{1.35}$$

$$g = -\left(K_2 + \beta C' H C\right)^{-1} \left(k_2 + \beta C' H b + \beta C' h\right) \tag{1.36}$$

$$H = K_1 + K_{12} G + \beta A' H (A + CG) \tag{1.37}$$

$$h = \left(K_{12} + \beta A' H C\right) g + k_1 + \beta A' (Hb + h). \tag{1.38}$$

Generally it is impossible to find the analytical solution to G, H, g, and h. Thus, an iterative procedure can designed as follows. First, set the initial H and h. G and g can then be calculated by (1.35) and (1.36). Given G and g as well as the initial H and h, the new H and h can be calculated by (1.37) and (1.38). Using the new H and h, one calculates the new G and g by (1.35) and (1.36) again. The process continues until convergence is achieved.[6]

In comparison with the log-linear approximation, Chow's method requires fewer derivations that must be accomplished by hand. Given the steady state, the only derivation is to obtain the partial derivatives of U and F. Even this can be computed with a written procedure in a major software package.[7] Despite this significant advantage, Chow's method has at least three weaknesses.

First, Chow's method can be a good approximation method only when the state equation is linear or can be transformed into a linear one. Otherwise, the linearized first-order condition as expressed by (1.31) and (1.32) will not be a good approximation to the nonapproximated (1.26) and (1.27), since A' and C' are not good approximations to $\frac{\partial F(x_t,u_t)}{\partial x_t}$ and $\frac{\partial F(x_t,u_t)}{\partial u_t}$.[8] Reiter (1997) has used a concrete example to show this point. Gong (1997) points out the same problem.

Second, the iteration with (1.35)–(1.38) may exhibit multiple solutions, because inserting (1.35) into (1.37) gives a quadratic matrix equation in H. Indeed, one would expect that the number of solutions will increase with the increase of the dimensions of state space.[9]

Third, the assumed state equation (1.25) is only a special case of (1.2). This will create some difficulty when being applied to the model with exogenous variables. One possible way to circumvent this problem is to regard the

exogenous variables, if there are any, as a part of state variables. This actually has been done by Chow and Kwan (1998) when the method is applied to a practical problem. Yet this will increase the dimension of state space and hence intensify the problem of multiple solutions.

1.5 An Algorithm for the Linear-Quadratic Approximation

In this section, we present an algorithm for solving a dynamic optimization model with the general formulation expressed in (1.1)–(1.3). The first-order condition used for this algorithm is derived from the Lagrangian, which indicates that the method does not require the assumption $\partial F/\partial x = 0$. The approximation method we use here is the linear-quadratic approximation, and therefore the method does not require log-linear-approximation—which, in many cases, must be accomplished by hand. Proposition 1 below allows us to save further derivations when applying the method of undetermined coefficients. Indeed, if we use an existing software procedure to compute the partial derivatives with respect to U and F, the only derivation in applying our algorithm is to derive the steady state. Therefore, our suggested algorithm takes full advantage of, yet overcomes all the limitations occurring in, Chow's method.

Since our state equation takes the form of (1.2) rather than (1.25), the first-order condition is established by (1.26) and (1.27). Evaluating the first-order condition along with (1.2) and (1.3) at their certainty equivalence form, we derive the steady states for x_t, u_t, λ_t, and z_t, which we denote respectively as $\bar{x}$, $\bar{u}$, $\bar{\lambda}$, and $\bar{z}$. Taking the first-order Taylor approximation around the steady state for (1.14), (1.15), and (1.2), we obtain

$$F_{11}x_t + F_{12}u_t + F_{13}E_t\lambda_{t+1} + F_{14}z_t + f_1 = \lambda_t, \tag{1.39}$$

$$F_{21}x_t + F_{22}u_t + F_{23}E_t\lambda_{t+1} + F_{24}z_t + f_2 = 0, \tag{1.40}$$

$$x_{t+1} = Ax_t + Cu_t + Wz_t + b, \tag{1.41}$$

where, in particular,

$$\begin{aligned}
&A = F_x && C = F_u\\
&b = F(\bar{x}, \bar{u}, \bar{z}) - F_x\bar{x} - F_z\bar{z} - F_u\bar{u} && W = F_z\\
&F_{11} = U_{xx} + \beta\bar{\lambda}F_{xx} && F_{12} = U_{xu} + \beta\bar{\lambda}F_{xu}\\
&F_{13} = \beta A' && F_{14} = \beta\bar{\lambda}F_{xz}\\
&f_1 = U_x + \beta A'\bar{\lambda} - F_{11}\bar{x} - F_{12}\bar{u} - F_{13}\bar{\lambda} - F_{14}\bar{z} && F_{21} = U_{ux} + \beta\bar{\lambda}F_{ux}\\
&F_{22} = U_{uu} + \beta\bar{\lambda}F_{uu} && F_{23} = \beta C'\\
&f_2 = U_u + \beta C'\bar{\lambda} - F_{21}\bar{x} - F_{22}\bar{u} - F_{23}\bar{\lambda} - F_{24}\bar{z} && F_{24} = \beta\bar{\lambda}F_{uz}.
\end{aligned}$$

Here we define U_x as $\partial U/\partial x$ and U_{xx} as $\partial^2 U/\partial x \partial x$, all to be evaluated at the steady state. The similarity is applied to U_u, U_{uu}, U_{ux}, U_{xu}, F_{xx}, F_{uu}, F_{xz}, F_{ux}, and F_{uz}. The objective is to find the linear decision rule and the Lagrangian function:

$$u_t = Gx_t + Dz_t + g \tag{1.42}$$

$$\lambda_{t+1} = Hx_{t+1} + Qz_{t+1} + h. \tag{1.43}$$

The following is the proposition regarding the solution for (1.42) and (1.43).

Proposition 1 *Assume u_t and λ_{t+1} follow (1.42) and (1.43), respectively. Then the solution of G, D, g, H, Q, and h satisfies*

$$G = M + NH \tag{1.44}$$

$$D = R + NQ \tag{1.45}$$

$$g = Nh + m \tag{1.46}$$

$$HCNH + H(A + CM) + F_{13}^{-1}(F_{12}N - I_m)H + F_{13}^{-1}(F_{11} + F_{12}M) = 0 \tag{1.47}$$

$$\left[F_{13}^{-1}(I - F_{12}N) - HCN\right]Q - QP = H(W + CR) + F_{13}^{-1}(F_{14} + F_{12}R) \tag{1.48}$$

$$h = \left[F_{13}^{-1}(I - F_{12}N) - HCN - I_m\right]^{-1}\left[H(Cm + b) + Qp + F_{13}^{-1}(f_1 + F_{12}m)\right] \tag{1.49}$$

where I_m is the $m \times m$ identity matrix and

$$N = \left(F_{23}F_{13}^{-1}F_{12} - F_{22}\right)^{-1} F_{23}F_{13}^{-1} \tag{1.50}$$

$$M = \left(F_{23}F_{13}^{-1}F_{12} - F_{22}\right)^{-1} (F_{21} - F_{23}F_{13}^{-1}F_{11}) \tag{1.51}$$

$$R = \left(F_{23}F_{13}^{-1}F_{12} - F_{22}\right)^{-1} (F_{24} - F_{23}F_{13}^{-1}F_{14}) \tag{1.52}$$

$$m = \left(F_{23}F_{13}^{-1}F_{12} - F_{22}\right)^{-1} (f_2 - F_{23}F_{13}^{-1}f_1). \tag{1.53}$$

Given H, the proposition allows us to solve G, Q, and h directly, according to (1.44), (1.48), and (1.49). Then D and g can be computed from (1.45) and (1.46). The solution to H is implied by (1.47), which is nonlinear (quadratic) in H. Obviously, if the model has more than one state variable, we cannot solve for H analytically. In this case, we rewrite (1.47) as

$$H = F(H), \tag{1.54}$$

iterating (1.54) until convergence gives us a solution to H.[10]

However, when one encounters the model with one state variable,[11] H becomes a scalar, and therefore (1.47) can be written as

$$a_1 H^2 + a_2 H + a_3 = 0,$$

with the two solutions given by

$$H_{1,2} = \frac{1}{2a_1}\left[-a_2 \pm \left(a_2^2 - 4a_1 a_3\right)^{\frac{1}{2}}\right].$$

In other words, the solutions can be computed without iteration. Further, in most cases, one can easily identify the proper solution by relying on the economic meaning of λ_t. For example, in all the models that we present in this book, the state equation is the capital stock. Therefore, λ_t is the shadow price of capital, which should be inversely related to the quantity of capital. This indicates that only the negative solution is a proper solution.

1.6 A Dynamic Programming Algorithm

In this section we describe a dynamic programming algorithm that enables us to compute optimal value functions as well as optimal trajectories of discounted optimal control problems of the type above. An extension to a stochastic decision problem is briefly summarized in appendix III.

The basic discretization procedure goes back to Capuzzo-Dolcetta (1983) and Falcone (1987), and is applied with an adaptive gridding strategy by Grüne (1997) and Grüne and Semmler (2004a). We consider discounted optimal control problems in discrete time $t \in \mathbb{N}_0$, given by

$$V(x) = \max_{u \in \mathcal{U}} \sum_{t=0}^{\infty} \beta^t g(x(t), u(t)), \tag{1.55}$$

where

$$x_{t+1} = f(x(t), u(t)), \ \ x(0) = x_0 \in \mathbb{R}^n.$$

For the discretization in space, we consider a grid Γ covering the computational domain of interest. Denoting the nodes of the grid Γ by x^i, $i = 1, \ldots, P$, we are now looking for an approximation V_h^Γ satisfying

$$V_h^\Gamma(x^i) = T_h\left(V_h^\Gamma\right)(x^i) \tag{1.56}$$

for all nodes x^i of the grid, where the value of V_h^Γ for points x that are not grid points (these are needed for the evaluation of T_h) is determined by linear interpolation. (For a description of several iterative methods for the solution of (1.56), see Grüne and Semmler 2004a).

For the estimation of the gridding error, we estimate the residual of the operator T_h with respect to V_h^Γ, that is, the difference between $V_h^\Gamma(x)$ and $T_h(V_h^\Gamma)(x)$ for points x that are not nodes of the grid. Thus, for each cell C_l of the grid Γ, we compute

$$\eta_l := \max_{x \in C_l} \left| T_h(V_h^\Gamma)(x) - V_h^\Gamma(x) \right|.$$

Using these estimates, we can iteratively construct adaptive grids as follows:

(0) Pick an initial grid Γ_0 and set $i = 0$. Fix a refinement parameter $\theta \in (0, 1)$ and a tolerance $tol > 0$.
(1) Compute the solution $V_h^{\Gamma_i}$ on Γ_i.
(2) Evaluate the error estimates η_l. If $\eta_l < tol$ for all l, stop.
(3) Refine all cells C_j with $\eta_j \geq \theta \max \eta_l$, set $i = i + 1$, and go to (1).

For more information about this adaptive gridding procedure and a comparison with other adaptive dynamic programming approaches, see Grüne and Semmler (2004a) and Grüne (1997).

In order to determine equilibria and approximately optimal trajectories, we need an approximately optimal policy, which in our discretization can be obtained in feedback form $u^*(x)$ for the discrete time approximation, using the following procedure.

For each x in the gridding domain, we choose $u^*(x)$ such that the equality

$$\max_{u \in U} \left\{ hg(x, u) + \beta V_h(x_h(1)) \right\} = \left\{ hg(x, u^*(x)) + \beta V_h(x_h^*(1)) \right\}$$

holds where $x_h^*(1) = x + hf(x, u^*(x))$. Then the resulting sequence $u_i^* = u^*(x_h(i))$ is an approximately optimal policy and the related piecewise constant control function is approximately optimal.

1.7 Conclusion

This chapter reviewed some typical approximation methods to solve a stochastic dynamic optimization model. The approximation methods discussed here used two types of first-order conditions: the Euler equation and the equations derived from the Lagrangian. The Euler equation needs a restriction that the state variable cannot appear as a determinant in the state equation. Although many economic models can satisfy this restriction after some transformation, we still cannot exclude the possibility that sometimes the restriction cannot be satisfied.

Given these two types of first-order conditions, we considered the solutions computed by the Fair–Taylor method, the log-linear approximation method, and the linear-quadratic approximation method. For all of these methods, we compared their advantages and disadvantages. We found that the Fair–Taylor method may encounter an accuracy problem due to its additional requirement of the initial condition for the control variable. On the other hand, the methods of log-linear approximation may need an algorithm that can take over some lengthy derivation process that otherwise must be analytically accomplished. For the linear-quadratic approximation, we therefore proposed an algorithm that could overcome the limitation of existing methods (such as Chow's method). We also elaborated on dynamic programming as a recently developed method to solve the Bellman equation. In chapter 2 we will turn to a practical problem and apply the various methods.

1.8 Appendix I: Proof of Proposition 1

From (1.39), we obtain

$$E_t\lambda_{t+1} = F_{13}^{-1}\left(\lambda_t - F_{11}x_t - F_{12}u_t - F_{14}z_t - f_1\right). \tag{1.57}$$

Substituting the above equation into (1.40) and then solving for u_t, we get

$$u_t = N\lambda_t + Mx_t + Rz_t + m, \tag{1.58}$$

where N, M, R, and m are all defined in the proposition. Substituting (1.58) into (1.41) and (1.55), we obtain, respectively,

$$x_{t+1} = S_x x_t + S_\lambda \lambda_t + S_z z_t + s \tag{1.59}$$

$$E_t\lambda_{t+1} = L_x x_t + L_\lambda \lambda_t + L_z z_t + l, \tag{1.60}$$

where

$$S_x = A + CM \tag{1.61}$$

$$S_\lambda = CN \tag{1.62}$$

$$S_z = W + CR \tag{1.63}$$

$$s = Cm + b \tag{1.64}$$

$$L_x = -F_{13}^{-1}(F_{11} + F_{12}M) \tag{1.65}$$

$$L_\lambda = F_{13}^{-1}(I - F_{12}N) \tag{1.66}$$

$$L_z = -F_{13}^{-1}(F_{14} + F_{12}R) \tag{1.67}$$

$$l = -F_{13}^{-1}(f_1 + F_{12}m). \tag{1.68}$$

Now express λ_t in terms of (1.43) and then plug it into (1.59) and (1.60):

$$x_{t+1} = (S_x + S_\lambda H)x_t + (S_z + S_\lambda Q)z_t + s + S_\lambda h, \tag{1.69}$$

$$E_t\lambda_{t+1} = (L_x + L_\lambda H)x_t + (L_z + L_\lambda Q)z_t + l + L_\lambda h. \tag{1.70}$$

Next, taking expectation for both sides of (1.43) and expressing x_{t+1} in terms of (1.41) and $E_t z_{t+1}$ in terms of (1.3), we obtain

$$E_t\lambda_{t+1} = HAx_t + HCu_t + (HW + QP)z_t + Hb + Qp + h. \tag{1.71}$$

Then, expressing u_t in terms of (1.42) for equations (1.41) and (1.71), we get, respectively,

$$x_{t+1} = (A + CG)x_t + (W + CD)z_t + Cg + b \tag{1.72}$$

$$E_t\lambda_{t+1} = H(A + CG)x_t + [H(W + CD) + QP]\, z_t + H(Cg + b) + Qp + h. \tag{1.73}$$

Comparing (1.69) and (1.70) with (1.72) and (1.73), we obtain

$$S_x + S_\lambda H = A + CG, \tag{1.74}$$

$$L_x + L_\lambda H = H(A + CG), \tag{1.75}$$

$$S_z + S_\lambda Q = W + CD, \tag{1.76}$$

$$L_z + L_\lambda Q = H(W + CD) + QP, \tag{1.77}$$

$$s + S_\lambda h = Cg + b, \tag{1.78}$$

$$l + L_\lambda h = H(Cg + b) + Qp + h. \tag{1.79}$$

These six equations determine six unknown coefficient matrices and vectors G, H, D, Q, g, and h. In particular, H is resolved from (1.75) when $A + CG$ is replaced by (1.74). This gives rise to (1.47) as in proposition 1, with S_x, S_λ, L_x, and L_λ to be expressed by (1.61), (1.62), (1.65), and (1.66), respectively. Given H, G is then resolved from (1.74), which allows us to obtain (1.44). Next, Q is resolved from (1.77) when $W+CD$ is replaced by (1.76). This gives rise to (1.48), with S_z, S_λ, L_z, and L_λ to be expressed by (1.63), (1.62), (1.67), and (1.66), respectively. Then D is resolved from (1.76), which allows us to obtain (1.45). Finally, h is resolved from (1.79) when $Cg + b$ is replaced by (1.78). This gives rise to (1.49), with S_λ, L_λ, s, and l to be expressed by (1.62), (1.66), (1.64), and (1.66), respectively. Finally g is resolved from (1.78), which allows us to obtain (1.46).

1.9 Appendix II: An Algorithm for the L-Q Approximation

The algorithm that we suggest for solving the L-Q approximation of section 1.5 is written as a GAUSS procedure and is available from the authors upon request. We call this procedure DYNPR. The input of this procedure includes the following:

- The steady states of x, u, λ, z, and F, denoted as *xbar*, *ubar*, *lbar*, *zbar*, and *Fbar*, respectively
- The first- and second-order partial derivatives of x, u, and z with respect to F and U, all evaluated at the steady state. They are denoted as, for instance, *Fx* and *Fxx* for the first- and second-order partial derivatives of x with respect to F, respectively
- The discount factor β (beta) and the parameters P and p (denoted as *BP* and *sp*, respectively), appearing in the $AR(1)$ process of z.

The output of this procedure is the decision parameters G, D, and g, denoted respectively as *BG*, *BD*, and *sg*.

PROC(3) = DYNPR $(Fx, Fu, Fz, Fxx, Fuu, Fzz, Fxu, Fxz, Fuz, Fux, Fzx, Ux, Uu, Uxx, Uuu, Uxu, Uux, Fbar, xbar, ubar, zbar, lbar, beta, BP, sp)$

LOCAL $A, C, W, sb, F11, F12, F13, F14, f1, F21, F22, F23, F24, f2, BM, BN, BR, sm, Sx, Slamda, Sz, ss, BLx, BLlamda, BLz, sl, sa1, sa2, sa3, BH, BG, sh, sg, BQ, BD$

$A = Fx$

$C = Fu$

$W = Fz$

$sb = Fbar - A * xbar - C * ubar - W * zbar$

$F11 = Uxx + beta * lbar * Fxx$

$F12 = Uxu + beta * lbar * Fxu$

$F13 = beta * Fx$

$F14 = beta * lbar * Fxz$

$f1 = Ux + beta * lbar * Fx - F11 * xbar - F12 * ubar - F13 * lbar - F14 * zbar$

$F21 = Uux + beta * lbar * Fux$

$F22 = Uuu + beta * lbar * Fuu$

$F23 = beta * (Fu')$

$F24 = beta * lbar * Fuz$

$f2 = Uu + beta * (Fu') * lbar - F21 * xbar - F22 * ubar - F23 * lbar - F24 * zbar$

$BM = \mathrm{INV}(F23 * \mathrm{INV}(F13) * F12 - F22) * (F21 - F23 * \mathrm{INV}(F13) * F11)$

```
        BN = INV(F23*INV(F13) * F12 - F22)*
             F23*INV(F13)
        BR = INV(F23*INV(F13) * F12 - F22)*
             (F24 - F23*INV(F13) * F14)
        sm = INV(F23*INV(F13) * F12 - F22)*
             (f2 - F23*INV(F13) * f1)
        Sx = A + C * BM
        Slamda = C * BN
        Sz = W + C * BR
        ss = C * sm + sb
        BLx = -INV(F13) * (F11 + F12 * BM)
        BLlamda = INV(F13) * (1 - F12 * BN)
        BLz = -INV(F13) * (F14 + F12 * BR)
        sl = -INV(F13) * (f1 + F12 * sm)
        sa1 = Slamda
        sa2 = Sx - BLlamda
        sa3 = -BLx
        BH = (1/(2 * sa1)) * (-sa2 - (sa2^2 - 4 * sa1 * sa3)^0.5)
        BG = BM + BN * BH
        BQ = INV(BLlamda - BH * Slamda - BP)*
             (BH * Sz - BLz)
        BD = BR + BN * BQ
        sh = INV(BLlamda - BH * Slamda - 1)*
             (BH * ss + BQ * sp - sl)
        sg = BN * sh + sm
  RETP(BG, BD, sg)
ENDP.
```

1.10 Appendix III: The Stochastic Dynamic Programming Algorithm

Our adaptive approach of section 1.6 is easily extended to stochastic discrete time problems of the type

$$V(x) = E\left(\max_{uin\ U} \sum_{t=0}^{\infty} \beta^t g(x(t), u(t))\right), \tag{1.80}$$

where

$$x(t+1) = f(x(t), u(t), z_t), \quad x(0) = x_0 \in R^n \tag{1.81}$$

and the z_t are i.i.d. random variables. This problem can immediately be applied in discrete time with the time step $h = 1$.[12]

The corresponding dynamic programming operator becomes

$$T_h(V_h)(x) = \max_{u \in U} E\left\{hg(x, u) + \beta V_h(\varphi(x, u, z))\right\}, \tag{1.82}$$

where $\varphi(x, u, z)$ is now a random variable.

If the random variable z is discrete, then the evaluation of the expectation E is a simple summation. If z is a continuous random variable, then we can compute E via a numerical quadrature formula for the approximation of the integral

$$\int_z (hg(x, u) + \beta V_h(\varphi(x, u, z))p(z)dz,$$

where $p(z)$ is the probability density of z. Grüne and Semmler (2004a) show the application of such a method to such a problem, where z is a truncated Gaussian random variable and the numerical integration is done via the trapezoidal rule.

Despite the formal similarity, stochastic optimal control problems have several features different from deterministic ones. First, complicated dynamical behavior, such as multiple, stable steady-state equilibria and periodic attractors, is less likely because the influence of the stochastic term tends to "smear out" the dynamics in such a way that these phenomena disappear.[13] Furthermore, in stochastic problems the optimal value function typically has more regularity, which allows the use of high-order approximation techniques. Finally, stochastic problems can often be formulated in terms of Markov decision problems with continuous transition probability (see Rust 1996 for details) whose structure gives rise to different approximation techniques, in particular allowing the avoidance of the discretization of the state space.

In these situations, the above dynamic programming technique, developed by Grüne (1997) and applied in Grüne and Semmler (2004a) may not be the most efficient approach to these problems. Nevertheless, the examples in Grüne and Semmler show that adaptive grids, discussed in section 1.6, are by far more efficient than nonadaptive methods if the same discretization technique is used for both. In the smooth case, one can obtain estimates for the error in the approximation of the gradient of V_h from our error estimates (for details, see Grüne 2003).

2

Solving a Prototype Stochastic Dynamic Model

2.1 Introduction

This chapter turns to a practical problem of dynamic optimization. In particular, we will solve a prototype model by employing the different approximation methods discussed in chapter 1. The model we chose is a Ramsey model (Ramsey 1928), to which the exact solution is computable with the standard recursive method. This will allow us to test the accuracy of approximations by comparing the different approximate solutions against the exact solution.

2.2 The Ramsey Problem

2.2.1 The Model

Ramsey (1928) posed a problem of optimal resource allocation that is now often used as a prototype model of dynamic optimization.[1] The model presented in this section is essentially that of Ramsey, but it is augmented by uncertainty. Let C_t denote consumption, Y_t output, and K_t the capital stock. Assume that the output is produced by capital stock and is either consumed or invested (that is, added to the capital stock). Formally,

$$Y_t = A_t K_t^{\alpha}, \tag{2.1}$$

$$Y_t = C_t + K_{t+1}, \tag{2.2}$$

where $\alpha \in (0, 1)$ and A_t is the technology that may follow an AR(1) process:

$$A_{t+1} = a_0 + a_1 A_t + \epsilon_{t+1}. \tag{2.3}$$

Here we will assume that ϵ_t is i.i.d. Equations (2.1) and (2.2) indicate that we can write the transition law of capital stock as

$$K_{t+1} = A_t K_t^{\alpha} - C_t. \tag{2.4}$$

We have assumed here that the depreciation rate of capital stock is equal to 1. This is a simplified assumption by which the exact solution is computable.[2]

The representative agent is assumed to find the control sequence $\{C_t\}_{t=0}^{\infty}$ such that

$$\max\ E_0 \sum_{t=0}^{\infty} \beta^t \ \ln\ C_t, \tag{2.5}$$

given the initial condition (K_0, A_0).

2.2.2 The Exact Solution and the Steady States

The exact solution for this model—which could be derived from the standard recursive method—can be written as

$$K_{t+1} = \alpha\beta A_t K_t^{\alpha}. \tag{2.6}$$

This further implies, from (2.4), that

$$C_t = (1 - \alpha\beta) A_t K_t^{\alpha}. \tag{2.7}$$

Given the solution paths for C_t and K_{t+1}, we are able to derive the steady state. It is not difficult to find that one steady state is on the boundary: $\bar{K} = 0$ and $\bar{C} = 0$. To obtain a more meaningful interior steady state, we take the logarithm for both sides of (2.6) and evaluate the equation at its certainty equivalence form:

$$\log K_{t+1} = \log(\alpha\beta\bar{A}) + \alpha \log K_t. \tag{2.8}$$

At the steady state, $K_{t+1} = K_t = \bar{K}$. Solving (2.8) for $\log\bar{K}$, we obtain $\log \bar{K} = \frac{\log(\alpha\beta\bar{A})}{1-\alpha}$. Therefore,

$$\bar{K} = (\alpha\beta\bar{A})^{1/(1-\alpha)}. \tag{2.9}$$

Given $\bar{K}$, $\bar{C}$ is resolved from (2.4):

$$\bar{C} = \bar{A}\bar{K}^{\alpha} - \bar{K}. \tag{2.10}$$

2.3 The First-Order Conditions and Approximate Solutions

To solve the model using different approximation methods, we will establish the first-order condition derived from the Ramsey problem. As mentioned in chapter 1, there are two types of first-order conditions, the Euler equation and the equation derived from the Lagrangian. Let us first consider the Euler equation.

2.3.1 The Euler Equation

Our first task is to transform the model into a setting such that the state variable K_t does not appear in $F(\cdot)$, as was discussed in chapter 1. This can be done by assuming K_{t+1} (instead of C_t) as the model's decision variable. To achieve a notational consistency in the time subscript, we denote the decision variable as Z_t. Therefore the model can be rewritten as

$$\max\ E_0 \sum_{t=0}^{\infty} \beta^t \ \ln\ \left(A_t K_t^{\alpha} - K_{t+1}\right), \tag{2.11}$$

subject to

$$K_{t+1} = Z_t.$$

Here we have used (2.4) to express C_t in the utility function. Also, in this formulation the state variable in period t is still K_t. Therefore, $\partial F/\partial x = 0$ and $\partial F/\partial u = 1$. The Bellman equation in this case can be written as

$$V(K_t, A_t) = \max_{K_{t+1}} \left\{\ln\ (A_t K_t^{\alpha} - K_{t+1}) + \beta E\left[V(K_{t+1}, A_{t+1})\right]\right\}. \tag{2.12}$$

The necessary condition for maximizing the right-hand side of the Bellman equation (2.12) is given by

$$\frac{-1}{A_t K_t^{\alpha} - K_{t+1}} + \beta E\left[\frac{\partial V}{\partial K_{t+1}}(K_{t+1}, A_{t+1})\right] = 0. \tag{2.13}$$

Meanwhile, applying the Benveniste–Scheinkman formula,

$$\frac{\partial V}{\partial K_t}(K_t, A_t) = \frac{\alpha A_t K_t^{\alpha-1}}{A_t K_t^{\alpha} - K_{t+1}}. \tag{2.14}$$

Substituting (2.14) into (2.13) allows us to obtain the Euler equation:

$$\frac{-1}{A_tK_t^\alpha - K_{t+1}} + \beta E\left[\frac{\alpha A_{t+1}K_{t+1}^{\alpha-1}}{A_{t+1}K_{t+1}^\alpha - K_{t+2}}\right] = 0,$$

which can be written as

$$-\frac{1}{C_t} + \beta E\left[\frac{\alpha A_{t+1}K_{t+1}^{\alpha-1}}{C_{t+1}}\right] = 0. \tag{2.15}$$

The Euler equations (2.15), (2.4), and (2.3) determine the transition sequences $\{K_{t+1}\}_{t=1}^{\infty}$, $\{A_{t+1}\}_{t=1}^{\infty}$, and $\{C_t\}_{t=0}^{\infty}$, given the initial conditions K_0 and A_0.

2.3.2 The First-Order Condition Derived from the Lagrangian

Next, we derive the first-order condition from the Lagrangian. Define the Lagrangian:

$$L = \sum_{t=0}^{\infty} \beta^t \ln C_t - \sum_{t=0}^{\infty} E_t\left[\beta^{t+1}\lambda_{t+1}\left(K_{t+1} - A_tK_t^\alpha + C_t\right)\right].$$

Setting to 0 the derivatives of L with respect to λ_t, C_t, and K_t, one obtains (2.4) as well as

$$1/C_t - \beta E_t\lambda_{t+1} = 0, \tag{2.16}$$

$$\beta E_t\lambda_{t+1}\alpha A_tK_t^{\alpha-1} = \lambda_t. \tag{2.17}$$

These are the first-order conditions derived from the Lagrangian. Next we try to demonstrate that the two first-order conditions are virtually equivalent. This can be done as follows. Using (2.16) to express $\beta E_t\lambda_{t+1}$ in terms of $1/C_t$ and then plugging it into (2.17), we obtain $\lambda_t = \alpha A_tK_t^{\alpha-1}/C_t$. This further indicates that

$$E_t\lambda_{t+1} = E_t\left[\alpha A_{t+1}K_{t+1}^{\alpha-1}/C_{t+1}\right] \tag{2.18}$$

Substituting (2.18) back into (2.16), we obtain the Euler equation (2.15).

It is not difficult to find that the two first-order conditions imply the same steady state as the one derived from the exact solution (see equations (2.9) and (2.10)). Writing either (2.15) or (2.17) in terms of the certainty equivalence form while evaluating them at the steady state, we indeed obtain (2.9).

2.3.3 The Dynamic Programming Formulation

The dynamic programming problem for the Ramsey growth model of section 2.2.1 can be written, in its deterministic version, as a basic discrete time growth model,

$$V = \max_{C} \sum_{t=0}^{\infty} \beta^t U(C_t) \tag{2.19}$$

s.t.

$$C_t + K_{t+1} = f(K_t) \tag{2.20}$$

with a one-period utility function $U'(C) > 0, U''(C) < 0, f'(K) > 0, f''(K) < 0$.

Let us restate the problem above with K as the state variable and K' as the control variable, where K' denotes the next period's value of K. Substitute C into the above intertemporal utility function by defining

$$C = f(K) - K'. \tag{2.21}$$

We then can express the discrete time Bellman equation, representing a dynamic programming formulation as

$$V(K) = \max_{K'} \{U[f(K) - K'] + \beta V(K')\}. \tag{2.22}$$

Applying the Benveniste–Scheinkman condition[3] gives

$$V'(K) = U'(f(K) - K')f'(K). \tag{2.23}$$

K is the state variable, and in equation (2.22) we have $V(K')$. From the discrete time form of the envelope condition, one again obtains the first-order condition of equation (2.22) as

$$-U'[f(K) - K'] + \beta V'(K') = 0,$$

which, by using (2.23) one step forward—for $V'(K')$—gives

$$U'[f(K) - K'] = \beta U'[f(K') - K'']f'(K'). \tag{2.24}$$

We thus obtain as a solution a second-order difference equation in K, whereby K' denotes the one-period-ahead and K'' the two-period-ahead value of K. However, equation (2.24) can be written as

$$1 = \beta \frac{U'(C_{t+1})}{U'(C_t)} f'(K_{t+1}), \tag{2.25}$$

which represents the Euler equation.[4]

If we allow for log-utility, as in section 2.2.1, the discrete time decision problem is directly analytically solvable. We take the following form of a utility function:

$$V = \max_{C_t} \sum_{t=0}^{\infty} \beta^t \ln C_t \tag{2.26}$$

s.t.

$$K_{t+1} = AK_t^{\alpha} - C_t. \tag{2.27}$$

The analytical solution for the value function is

$$V(K) = \tilde{B} + \tilde{C} \ln (K), \tag{2.28}$$

and for the sequence of capital, one obtains

$$K_{t+1} = \frac{\beta \tilde{C} A K_t^{\alpha}}{1 + \beta \tilde{C}}, \tag{2.29}$$

with

$$\tilde{C} = \frac{\alpha}{1 - \alpha\beta}$$

and

$$\tilde{B} = \frac{\ln (\tilde{C}(1 - \alpha\beta)A) + \frac{\beta\alpha}{1-\beta\alpha} \ln (\alpha\beta A)}{1 - \beta}.$$

For the optimal consumption,

$$C_t = K_{t+1} - AK_t^{\alpha}, \tag{2.30}$$

and for the steady-state equilibrium $\bar{K}$, one obtains

$$\frac{1}{\beta} = \alpha A \bar{K}^{\alpha - 1} \tag{2.31}$$

or

$$\bar{K} = \beta\alpha A \bar{K}^{\alpha}. \tag{2.32}$$

2.4 Solving the Ramsey Problem with Different Approximations

2.4.1 The Fair–Taylor Solution

One can apply the Fair–Taylor method either to the Euler equation or to the first-order condition derived from the Lagrangian. Here we will use the Euler equation. Let us first write equation (2.15) in the form expressed by equations (1.19)–(1.21):

$$C_{t+1} = \alpha\beta C_t A_{t+1} K_{t+1}^{\alpha - 1}. \tag{2.33}$$

Table 2.1. Parameterizing the Prototype Model

α	β	a_0	a_1	σ_ϵ	$\bar{K}$	$\bar{C}$	$\bar{A}$
0.3200	0.9800	600	0.8000	60.000	23,593	51,640	3,000

Together with (2.4) and (2.3), they form a recursive dynamic system from which the transition paths of C_t, K_t, and A_t can be directly computed. Since the model is simple in its structure, there is no necessity to employ the Gauss–Seidel procedure (as suggested in chapter 1).

Before we compute the solution path, we will parameterize the model. All together there are five structural parameters: α, β, a_0, a_1, and σ_ϵ. Table 2.1 specifies these parameters and the corresponding interior steady-state values.

Given the parameters in table 2.1, we provide in figure 2.1 three solution paths computed by the Fair–Taylor method. These solution paths are compared with the exact solution expressed in equations (2.6) and (2.7).

The three solution paths are different due to their initial conditions with regard to C_0. Since we know the exact solution, we thus can choose C_0 close to the exact solution, denoted as C_0^*. From (2.7), $C_0^* = (1 - \alpha\beta)A_0K_0^\alpha$. In particular, we allow one C_0 to be equal to C_0^*, and the others to deviate by 1% from C_0^*.

The following is a summary of what we found in this experiment.

- When we choose C_0 above C_0^* (by 1%) the path of K_t quickly reaches 0, and therefore the simulations must be subject to the constraint $C_t < A_tK_t^\alpha$. In particular, we restrict $C_t \leq 0.99A_tK_t^\alpha$. This restriction causes K_t never to reach 0 so that the simulation can be continued. The solution path is shown by the dashed curves in figure 2.1.
- When we set C_0 below C_0^* (again by 1%), the path of C_t quickly reaches its lower bound 0. This is shown by the dotted curve in figure 2.1 (right).
- When we set C_0 to C_0^*, the paths of K_t and C_t (shown by another dashed curve) are close to the exact solution for small t's. When t goes beyond a certain point, the deviation becomes significant (left).

What can we learn from this experiment? The exact solution to this problem seems to be the saddle path for the system composed of (2.4), (2.3), and (2.33). The eventual deviation of the solution starting with C_0^* from the exact solution is likely to be due to computational errors resulting from our numerical simulation. On the other hand, we have verified our previous concern that the initial condition for the control variable is extremely important for obtaining an appropriate solution path when we employ the Fair–Taylor method.

2.4.2 The Log-Linear Solution

As with the Fair–Taylor method, the log-linear approximation method can be applied to the first-order condition either from the Euler equation or

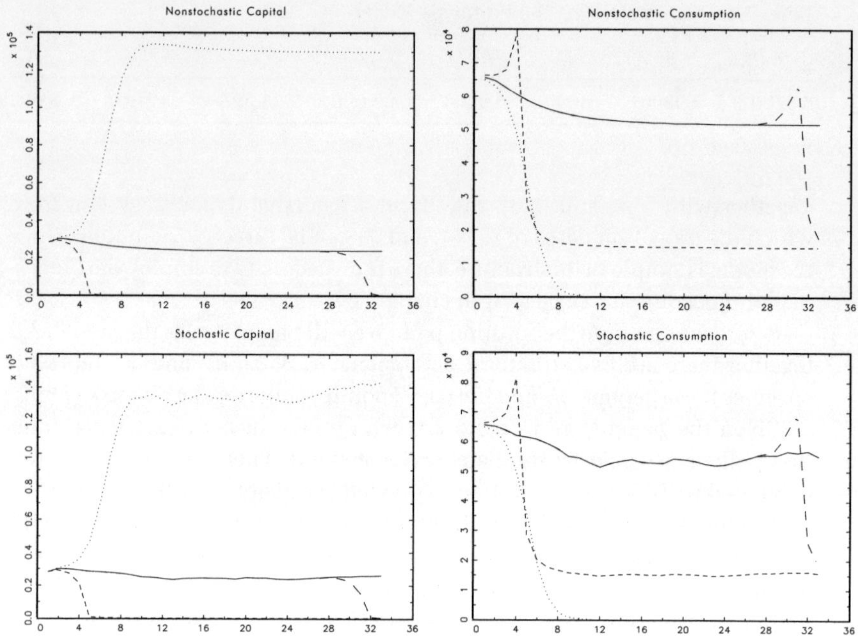

Figure 2.1. The Fair–Taylor solution compared with the exact solution. Note: Solid curve, exact solution; dashed and dotted curves, Fair–Taylor solution.

from the Lagrangian. Here we will again use the Euler equation. Our first task is therefore to log-linearize the state, Euler, and exogenous equations as expressed in (2.4), (2.15), and (2.3). The following is the proposition regarding this log-linearization (the proof is provided in appendix I).

Proposition 2 *Let k_t, c_t, and a_t denote the log deviations of K_t, C_t, and A_t. Then equations (2.4), (2.15), and (2.3) can be log-linearized as*

$$k_{t+1} = \varphi_{ka}a_t + \varphi_{kk}k_t + \varphi_{kc}c_t, \tag{2.34}$$

$$E\left[c_{t+1}\right] = \varphi_{cc}c_t + \varphi_{ca}a_t + \varphi_{ck}k_{t+1}, \tag{2.35}$$

$$E\left[a_{t+1}\right] = a_1 a_t, \tag{2.36}$$

where

$$\varphi_{ka} = \bar{A}\bar{K}^{\alpha-1}, \qquad \varphi_{kk} = \bar{A}\bar{K}^{\alpha-1}\alpha,$$
$$\varphi_{kc} = -(\bar{C}/\bar{K}), \qquad \varphi_{cc} = \alpha\beta\bar{A}\bar{K}^{\alpha-1},$$
$$\varphi_{ca} = \alpha\beta\bar{A}\bar{K}^{\alpha-1}a_1, \quad \varphi_{ck} = \alpha\beta\bar{A}\bar{K}^{\alpha-1}(\alpha-1).$$

Next we try to find a solution path for c_t, which we conjecture as

$$c_t = \eta_{ca} a_t + \eta_{ck} k_t. \tag{2.37}$$

The proposition below regards the determination of the undetermined coefficients η_c and η_{ck}.

Proposition 3 *Assume c_t follows (2.37). Then η_{ck} and η_{ca} are determined from the following:*

$$\eta_{ck} = \frac{1}{2Q_2}\left(-Q_1 - \sqrt{Q_1^2 - 4Q_0Q_2}\right) \tag{2.38}$$

$$\eta_{ca} = \frac{(\eta_{ck} - \varphi_{ck})\varphi_{ka} - \varphi_{ca}}{\varphi_{cc} - a_1 - \varphi_{kc}(\eta_{ck} - \varphi_{ck})}, \tag{2.39}$$

where $Q_2 = \varphi_{kc}$, $Q_1 = (\varphi_{kk} - \varphi_{cc} - \varphi_{kc}\varphi_{ck})$, and $Q_0 = \varphi_{kk}$.

The solution paths of the model can now be computed by relying on (2.37) and (2.34), with a_t given by $a_1 a_{t-1} + \epsilon_t/\bar{A}$. All the solution paths are expressed as log deviations. Therefore, to compare the log-linear solution with the exact solution, we perform the transformation via $X_t = (1 + x_t)\bar{X}$ for a variable x_t in log deviation form. Using the same parameters as in table 2.1, we show in figure 2.2 the log-linear solution in comparison with the exact solution.

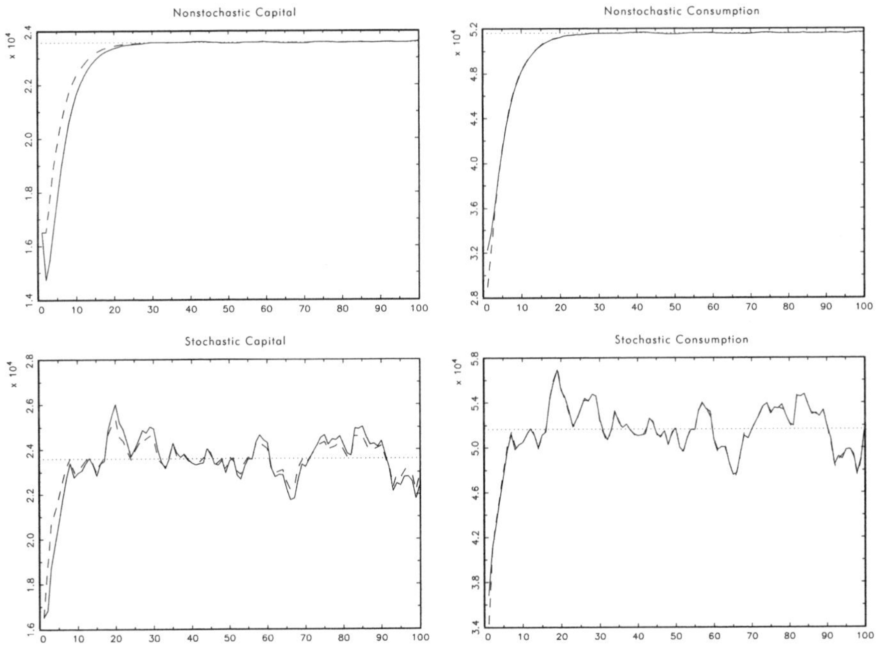

Figure 2.2. The log-linear solution compared with the exact solution. Note: Solid curve, exact solution; dashed curves, log-linear solution.

In contrast to the Fair–Taylor solution, the log-linear solution is quite close to the exact solution except for some initial paths.

2.4.3 The Linear-Quadratic Solution with Chow's Algorithm

To apply Chow's method, we first transform the model in which the state equation appears to be linear. This can be done by choosing investment $I_t \equiv A_t K_t^\alpha - C_t$ as the control variable while leaving capital stock and technology as the two state variables. When considering this, the model becomes

$$\max E_0 \sum_{t=0}^{\infty} \beta^t \ln (A_t K_t^\alpha - I_t),$$

subject to (2.3) as well as

$$K_{t+1} = I_t. \tag{2.40}$$

Due to the insufficiency in the specification of the model with regard to the possible exogenous variable, we also have to treat, as suggested in chapter 1, the technology A_t as a state variable. This indicates that there are two state equations, (2.40) and (2.3), and both are now in linear form. Next we derive the first- and second-order partial derivatives of the utility function around the steady state. This will allow us to obtain the K_{ij} and k_j $(i, j = 1, 2)$ coefficient matrices and vectors expressed in Chow's first-order conditions (1.31) and (1.32).

Suppose the linear decision rule can be written as

$$I_t = G_{11} K_t + G_{12} A_t + g_1. \tag{2.41}$$

The coefficients G_{11}, G_{12}, and g_1 can be computed in principle by iterating equations (1.35)–(1.38), as discussed in chapter 1. However, this requires the iteration to be convergent. Unfortunately, this is not attainable for our particular application, even if we start from many different initial conditions.[5] Therefore, our attempt to compute the solution path with Chow's algorithm fails.

2.4.4 The Linear-Quadratic Solution Using the Suggested Algorithm

When we employ our new algorithm, there is no need to transform the model. Therefore we can define $F = AK^\alpha - C$ and $U = \ln C$. Again our first step is to compute the first- and second-order partial derivatives with respect to F and U. All these partial derivatives, along with the steady states, can be used as

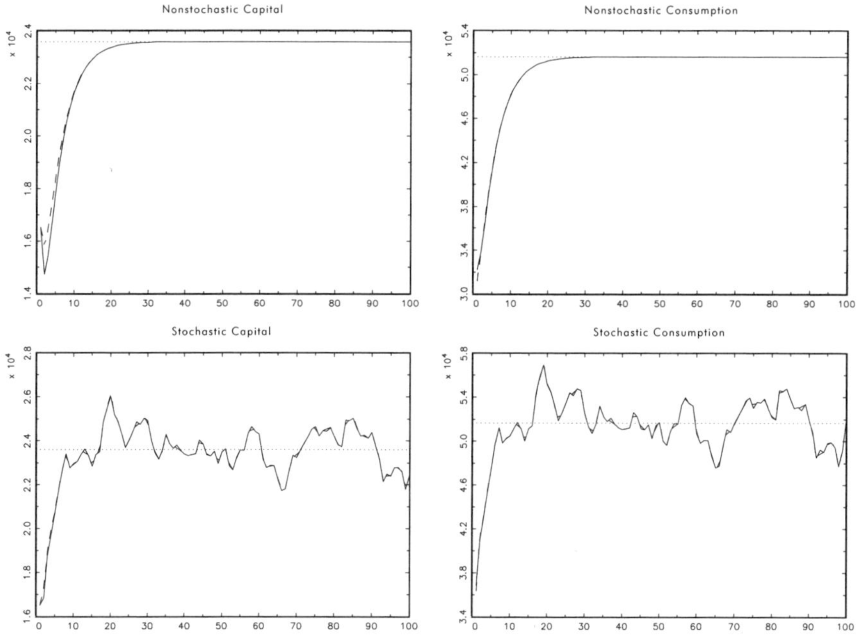

Figure 2.3. The linear-quadratic solution compared with the exact solution. Note: Solid curve, exact solution; dashed curves, linear-quadratic solution.

input in the GAUSS procedure provided in appendix II of chapter 1. Executing this procedure will allow us to compute the undetermined coefficients in the following decision rule for C_t:

$$C_t = G_{21}K_t + G_{22}A_t + g_2. \tag{2.42}$$

Equations (2.42), (2.4), and (2.3) form the dynamic system from which the transition paths of C_t, K_t, and A_t are computed (see figure 2.3 for an illustration).

In addition, in figure 2.4, the value function obtained from the linear quadratic solution is shown. The value function is clearly concave in the capital stock, K.

2.4.5 The Dynamic Programming Solution

Next, we compare the analytical solution of section 2.3.3 with the dynamic programming solution obtained from the dynamic programming algorithm of chapter 1, section 1.6. We will report only results from a deterministic version. Results from a stochastic version are discussed in appendix II.

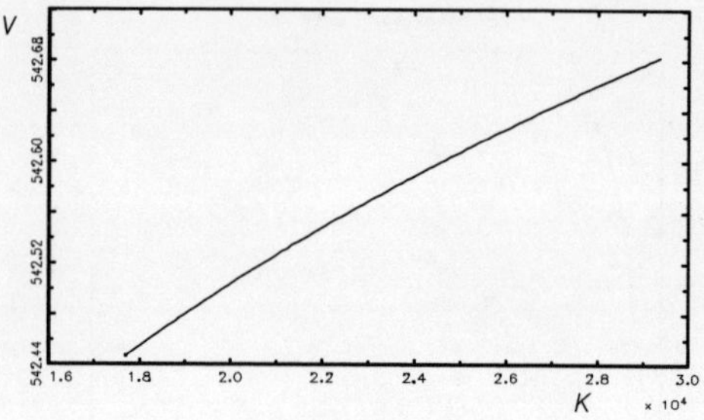

Figure 2.4. Value Function Obtained from the Linear-Quadratic Solution

For the growth model of section 2.3.3, we employ the parameters:

$$\alpha = 0.34$$

$$A = 5$$

$$\beta = 0.95.$$

We can solve all the above expressions numerically for a grid of the capital stock in the interval [0.1, 10] and the control variable, c, in the interval [0.1, 5]. For the parameters chosen, we obtain a steady state of the capital stock:

$$\bar{K} = 2.07.$$

For more details of the solution, see Grüne and Semmler (2004a).[6]

The solution of the growth model with the above parameters, using the dynamic programming algorithm of chapter 1, section 1.6, with grid refinement, is shown in figures 2.5 and 2.6.

As figures 2.5 and 2.6 show, the value function and the control, C, are concave in the capital stock, K, in figure 2.6. The optimal consumption is shown to depend on the state variable K for a grid of K, $0 \leq K \leq 10$. Moreover, as can be seen in figure 2.6, consumption is low when capital stock is low (capital stock can grow) and consumption is high when capital stock is high (capital stock will decrease), where "low" and "high" refer to the optimal steady-state capital stock $\bar{K} = 2.07$. As reported in Grüne

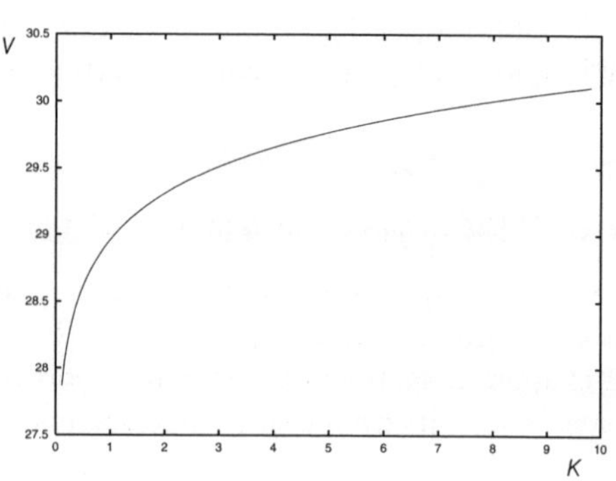

Figure 2.5. Value Function

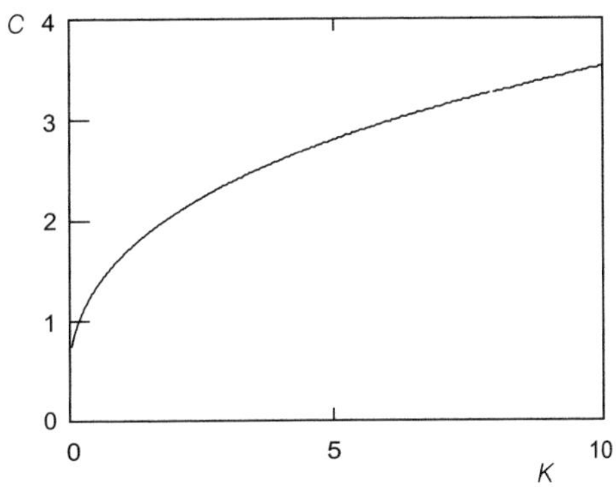

Figure 2.6.
Path of the Choice Variable C

and Semmler (2004a), the dynamic programming algorithm with adaptive gridding strategy, as introduced in section 1.6, solves the value function with high accuracy.[7]

2.5 Conclusion

This chapter employed the various approximation methods to solve a prototype dynamic optimization model. Our purpose here was to compare the different approximate solutions with the exact solution, which for this model can be derived analytically by the standard recursive method. There were some difficulties when we applied the Fair–Taylor method and the method of linear-quadratic approximation using Chow's algorithm. However, when we applied log-linear approximation and linear-quadratic approximation with our suggested algorithm, the approximate solutions were close to the exact solution. We also found that the method of log-linear approximation may need an algorithm that can take over some lengthy derivations that otherwise must be analytically accomplished. Therefore, our experiment in this chapter verified our concerns with regard to the accuracy and the capability of different approximation methods, including the Fair–Taylor method, the log-linear approximation method, and the linear-quadratic approximation method. Although the dynamic programming approach solved the value function with higher accuracy, in subsequent chapters, when we come to the calibration of the intertemporal decision models, we will work with the linear-quadratic approximation of the Chow method because it is more applicable to empirical assessment of the models.

2.6 Appendix I: The Proof of Propositions 2 and 3

2.6.1 The Proof of Proposition 2

For convenience, we write (2.4), (2.15), and (2.3) as

$$K_{t+1} - A_t K_t^{\alpha} - C_t = 0 \tag{2.43}$$

$$E\left[C_{t+1} - \alpha\beta C_t A_{t+1} K_{t+1}^{\alpha-1}\right] = 0 \tag{2.44}$$

$$E\left[A_{t+1} - a_0 - a_1 A_t\right] = 0. \tag{2.45}$$

Applying (1.22) to the above equations, we obtain

$$\bar{K}e^{k_{t+1}} - \bar{A}\bar{K}^{\alpha}e^{a_t+\alpha k_t} + \bar{C}e^{c_t} = 0 \tag{2.46}$$

$$E\left[\bar{C}e^{c_{t+1}} - \alpha\beta\bar{C}\bar{A}\bar{K}^{\alpha-1}e^{a_{t+1}+c_t+(\alpha-1)k_{t+1}}\right] = 0 \tag{2.47}$$

$$E\left[\bar{A}e^{a_{t+1}} - a_0 - a_1\bar{A}e^{a_t}\right] = 0. \tag{2.48}$$

Applying (1.23), from the above we obtain

$$\bar{K}(1 + k_{t+1}) - \bar{A}\bar{K}^{\alpha}(1 + a_t + \alpha k_t) + \bar{C}(1 + c_t) = 0$$

$$E\left[\bar{C}(1 + c_{t+1}) - \alpha\beta\bar{C}\bar{A}\bar{K}^{\alpha-1}(1 + c_t + a_{t+1} + (\alpha - 1)k_{t+1})\right] = 0$$

$$E\left[\bar{A}(1 + a_{t+1}) - a_0 - a_1\bar{A}(1 + a_t)\right] = 0,$$

which can be written as

$$\bar{K}k_{t+1} - \bar{A}\bar{K}^{\alpha}a_t - \bar{A}\bar{K}^{\alpha}\alpha k_t + \bar{C}c_t = 0 \tag{2.49}$$

$$E\left[\bar{C}c_{t+1} - \alpha\beta\bar{C}\bar{A}\bar{K}^{\alpha-1}(c_t + a_{t+1} + (\alpha - 1)k_{t+1})\right] = 0 \tag{2.50}$$

$$E\left[a_{t+1} - a_1 a_t\right] = 0. \tag{2.51}$$

Equation (2.51) indicates (2.36). Substituting it into (2.50) to express $E\left[a_{t+1}\right]$ and rearranging (2.49) and (2.50), we obtain (2.34) and (2.35), as indicated in the proposition.

2.6.2 Proof of Proposition 3

Given the conjectured solution (2.37), the transition path of k_{t+1} can be derived from (2.34) and can be written as

$$k_{t+1} = \eta_{ka}a_t + \eta_{kk}k_t, \tag{2.52}$$

where

$$\eta_{ka} = \varphi_{ka} + \varphi_{kc}\eta_{ca} \tag{2.53}$$

$$\eta_{kk} = \varphi_{kk} + \varphi_{kc}\eta_{ck}. \tag{2.54}$$

Expressing c_{t+1} and c_t in terms of (2.37) while recognizing that $E\left[a_{t+1}\right] = a_1 a_t$, from (2.35) we obtain:

$$\eta_{ca}a_1 a_t + \eta_{ck}k_{t+1} = \varphi_{cc}(\eta_{ca}a_t + \eta_{ck}k_t) + \varphi_{ca}a_t + \varphi_{ck}k_{t+1},$$

which can further be written as

$$k_{t+1} = \frac{\varphi_{cc}\eta_{ca} + \varphi_{ca} - \eta_{ca}a_1}{\eta_{ck} - \varphi_{ck}} a_t + \frac{\varphi_{cc}\eta_{ck}}{\eta_{ck} - \varphi_{ck}} k_t. \tag{2.55}$$

Comparing (2.55) with (2.52), with η_{ka} and η_{kk} to be given by (2.53) and (2.54), we obtain

$$\frac{\varphi_{cc}\eta_{ca} + \varphi_{ca} - \eta_{ca}a_1}{\eta_{ck} - \varphi_{ck}} = \varphi_{ka} + \varphi_{kc}\eta_{ca} \tag{2.56}$$

$$\frac{\varphi_{cc}\eta_{ck}}{(\eta_{ck} - \varphi_{ck})} = \varphi_{kk} + \varphi_{kc}\eta_{ck}. \tag{2.57}$$

Equation (2.57) gives rise to the following quadratic function in η_{ck}:

$$Q_2\eta_{ck}^2 + Q_1\eta_{ck} + Q_0 = 0, \tag{2.58}$$

with Q_2, Q_1, and Q_0 to be given in the proposition. Solving (2.58) for η_{ck}, we obtain (2.38). Given η_{ck}, η_{ca} is resolved from (2.56), which gives rise to (2.39).

2.7 Appendix II: Dynamic Programming for the Stochastic Version

Here we present a stochastic version of a growth model that is based on the Ramsey model of section 2.1 but extended to the stochastic case. A model of the type goes back to Brock and Mirman (1972). Here the Ramsey 1d model of section 2.3.3 is extended, using a second variable modeling a stochastic shock. The model is given by the discrete time equations

$$K(t+1) = A(t)\tilde{A}K(t)^{\alpha} - C(t)$$

$$A(t+1) = \exp(\rho \ln A(t) + z_t),$$

where α and ρ are real constants and the $z(t)$ are i.i.d. random variables with 0 mean. The return function is again $U(C) = \ln C$.

In our numerical computations, which follow Grüne and Semmler (2004a), we used the parameter values $\tilde{A} = 5, \alpha = 0.34, \rho = 0.9$, and $\beta = 0.95$. As in the case of the Ramsey model, the exact solution is known and given by

$$V(K, A) = B + \tilde{C} \ln K + DA,$$

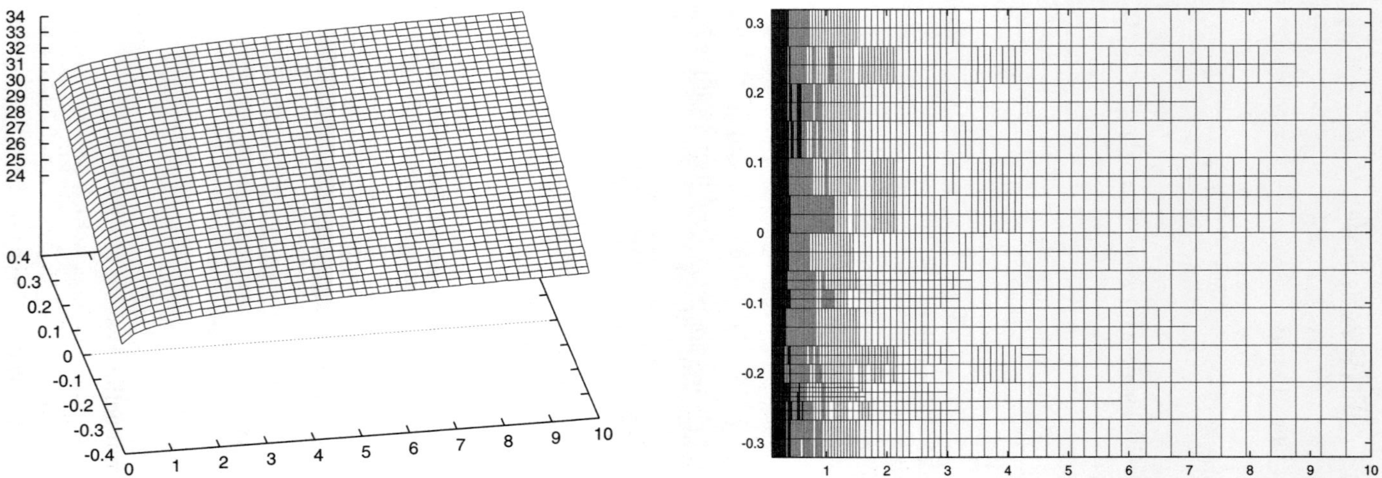

Figure 2.7. Approximated Value Function and Final Adaptive Grid for Our Example

Table 2.2. Number of Nodes and Errors for Our Example

No. Nodes	Error	Estimated Error
49	$1.4 \cdot 10^0$	$1.6 \cdot 10^1$
56	$0.5 \cdot 10^{-1}$	$6.9 \cdot 10^0$
65	$2.9 \cdot 10^{-1}$	$3.4 \cdot 10^0$
109	$1.3 \cdot 10^{-1}$	$1.6 \cdot 10^0$
154	$5.5 \cdot 10^{-2}$	$6.8 \cdot 10^{-1}$
327	$2.2 \cdot 10^{-2}$	$2.4 \cdot 10^{-1}$
889	$9.6 \cdot 10^{-3}$	$7.3 \cdot 10^{-2}$
2,977	$4.3 \cdot 10^{-3}$	$3.2 \cdot 10^{-2}$

where

$$B = \frac{\ln((1-\beta\alpha)\tilde{A}) + \frac{\beta\alpha}{1-\beta\alpha}\ln(\beta\alpha\tilde{A})}{1-\beta}, \quad \tilde{C} = \frac{\alpha}{1-\alpha\beta}, \quad D = \frac{1}{(1-\alpha\beta)(1-\rho\beta)}.$$

We have computed the solution to this problem on the domain $\Omega = [0.1, 10] \times [-0.32, 0.32]$. The integral over the Gaussian variable z was approximated by a trapezoidal rule with eleven discrete values equidistributed in the interval $[-0.032, 0.032]$, which ensures $\varphi(x, u, z) \in \Omega$ for $x \in \Omega$ and suitable $u \in U = [0.5, 10.5]$. For evaluating the maximum in T, the set U was discretized with 161 points. Table 2.2 shows the results of the resulting adaptive gridding scheme applied with refinement threshold $\theta = 0.1$ and coarsening tolerance $ctol = 0.001$. Figure 2.7 shows the resulting optimal value function and adapted grid.

In Santos and Vigo-Aguiar (1998), on equidistant grids with $143 \times 9 = 1{,}287$ and $500 \times 33 = 16{,}500$ nodes, errors of $2.1 \cdot 10^{-1}$ and $1.48 \cdot 10^{-2}$, respectively, were reported. In our adaptive iteration, these accuracies could be obtained with 109 and 889 nodes, respectively; thus we obtain a reduction in the number of nodes of more than 90% in the first case and almost 95% in the second case, even though the anisotropy of the value function was already taken into account in these equidistant grids. Here, again, in our stochastic version of the growth model, a steep value function can best be approximated with grid refinement.

3

Estimating and Evaluating the Stochastic Dynamic Model

3.1 Introduction

Solving a stochastic dynamic optimization model with aproximation methods or dynamic programming is only a first step toward the empirical assessment of such a model. Another necessary step is to estimate the model with econometric techniques. To undertake this step, certain approximation methods are more useful than others. Given the estimation, one can then evaluate the model to see how its prediction matches the empirical data.

The task of estimation has often been ignored in the current empirical studies of a stochastic dynamic optimization model when a technique, often referred to as calibration, is employed. The calibration approach compares the moment statistics (usually the second moments) of major macroeconomic time series with those obtained from simulating the model.[1] Typically the parameters employed for the model's simulation are selected from independent sources, such as different microeconomic studies. This approach has been criticized because the structural parameters are assumed to be given rather than estimated.[2]

Although this may not create severe difficulties for some currently used stochastic dynamic optimization models, such as the RBC, we believe the problem remains for more elaborated models in the future development of macroeconomic research. Unfortunately, there are not many econometric studies on how to estimate a stochastic dynamic optimization model except for a few attempts that have been undertaken for some simple cases.[3] In this chapter, we discuss two estimation methods: the Generalized Method of Moment (GMM) and the Maximum Likelihood (ML) estimation. Both methods define an objective function to be optimized. Because of the complexity of stochastic dynamic models, it is often unclear how the parameters to be estimated are related to the model's restrictions and, hence, to the objective function in the estimation. We thus also need to develop an estimation strategy that can be used to recursively search the parameter space in order to obtain the optimum.

Section 3.2 introduces the calibration technique, which has been used in the current empirical study of stochastic dynamic models. A proper application of calibration requires us to define the model's structural parameters correctly. Section 3.3 considers two possible estimation methods, the GMM and the ML. Section 3.4 proposes a strategy to implement these two methods for estimating a dynamic optimization model. This estimation requires a global optimization algorithm. Therefore, section 3.5 introduces a global optimization algorithm, called simulated annealing, that is used for executing the suggested strategy of estimation. Finally, a sketch of the computer program for our estimation strategy is presented in the appendix.

3.2 Calibration

The current empirical studies of a stochastic dynamic model often rely on calibration. This approach uses the Monte Carlo method to generate the distribution of some moment statistics implied by the model. For an empirical assessment of the model, solved through an approximation method, we compare these moment statistics with the sample moments computed from the data. Generally, the calibration includes the following steps:

- Step 1. Select the model's structural parameters. These may include preference and technology parameters and those that describe the distribution of the random variables in the model, such as σ_ϵ, the standard deviation of ϵ_t in equation (1.3).
- Step 2. Select the number of times the iteration is conducted in the simulation of the model. This number might be the same as the number of observations in the sample. We denote this number by T.
- Step 3. Select the initial condition (x_0, z_0) and use the state equation (1.2), the control equation (1.4), and the exogenous equation (1.3) to compute the solution of the model iteratively T times. This can be regarded as a one-time simulation.
- Step 4. If necessary, detrend the simulated series generated in step 3 to remove its time trend. Often the HP-filter (see Hodrick and Prescott 1980) is used for this detrending.
- Step 5. Compute the moment statistics of interest using, if necessary, a detrended series generated in step 4. These moment statistics are mainly those of the second moments, such as variances and covariances.

- Step 6. Repeat steps 3–5 N times. Here N should be sufficiently large. Then, after these N repeated runs, compute the distributions of these moment statistics. These distributions are mainly represented by their means and their standard deviations.
- Step 7. Compute the same moment statistics from the data sample and check whether they fall within the proper range of the distribution for the moment statistics generated from the Monte Carlo simulation of the model.

Because of the stochastic innovation of ϵ_t, the simulated series should also be cyclically and stochastically fluctuating. The extent of fluctuations, reflected by second-moment statistics, should depend on the model specification and the structural parameters, including σ_ϵ, the standard deviation of ϵ_t. Therefore, if we specify the model correctly with the structural parameters—including σ_ϵ, which also should be defined properly—the above comparison of moment statistics of the model and sample should provide a basis to test whether the model can explain the actual business cycles represented by the data.

3.3 The Estimation Methods

The application of the calibration requires techniques to select the structural parameters accurately. This indicates that we need to estimate the model before the calibration. We consider two possible estimation methods: the GMM and the ML.

3.3.1 The Generalized Method of Moments (GMM) Estimation

The GMM estimation starts with a set of orthogonal conditions representing the population moments established by the theoretical model:

$$E\left[h(y_t, \psi)\right] = 0, \tag{3.1}$$

where y_t is a k-dimensional vector of observed random variables at date t; ψ is an l-dimensional vector of unknown parameters that needs to be estimated; $h(\cdot)$ is a vector-valued function mapping from $R^k \times R^l$ into R^m. Let y_T contain all the observations of the k variables in the sample with size T. The sample average of $h(\cdot)$ can then be written as

$$g_T(\psi; y_T) = \frac{1}{T}\sum_{t=1}^{T} h(y_t, \psi). \tag{3.2}$$

Note that $g_T(\cdot)$ is also a vector-valued function with m dimensions. The idea behind the GMM estimation is to choose an estimator of ψ, denoted

$\widehat{\psi}$, such that the sample moments $g_T(\widehat{\psi}; y_T)$ are as close as possible to the population moments reflected by (3.1). To achieve this, one needs to define a distance function by which that closeness can be judged. Hansen (1982) suggested the following distance function:

$$J(\widehat{\psi}; y_T) = \left[g_T(\widehat{\psi}; y_T)\right]' W_T \left[g_T(\widehat{\psi}; y_T)\right], \tag{3.3}$$

where W_T, the weighting matrix, is $m \times m$, symmetric, positive definite, and depends only on the sample observation y_T. The choice of this weighting matrix defines a metric that makes the distance function a scalar. The GMM estimator of ψ is the value of $\widehat{\psi}$ that minimizes (3.3). Hansen proves that under certain assumptions such a GMM estimator is consistent and asymptotically normal. Also from the results established in Hansen, a consistent estimator of the variance-covariance matrix of $\widehat{\psi}$ is given by

$$Var(\widehat{\psi}) = \frac{1}{T}(D_T)^{-1} W_T (D_T')^{-1}, \tag{3.4}$$

where $D_T = \partial g_T(\widehat{\psi})/\partial \psi'$.

There is great flexibility in the choice of W_T for constructing a consistent and asymptotically normal GMM estimator. We adopt the method of Newey and West (1987), who suggested that

$$W_T^{-1} = \widehat{\Omega}_0 + \sum_{j=1}^{d} w(j, d)(\widehat{\Omega}_j + \widehat{\Omega}_j'), \tag{3.5}$$

with $w(j, d) \equiv 1 - j/(1+d)$, $\widehat{\Omega}_j \equiv (1/T)\sum_{t=j+1}^{T} g(y_t, \widehat{\psi}^*) g(y_{t-j}, \widehat{\psi}^*)$, and d a suitable function of T. Here $\widehat{\psi}^*$ is required to be a consistent estimator of ψ. Therefore, the estimation with the GMM method usually requires two steps, as suggested by Hansen and Singleton (1982). First, one chooses a suboptimal weighting matrix to minimize (3.3), and hence obtains a consistent estimator $\widehat{\psi}^*$. Second, one then uses the consistent estimator to calculate the optimum W_T through which (3.3) is reminimized.

3.3.2 The Maximum Likelihood (ML) Estimation

The ML estimation, proposed by Chow (1993a) for estimating a dynamic optimization model, starts with an econometric model:

$$By_t + \Gamma x_t = \epsilon_t, \tag{3.6}$$

where B is a $m \times m$ matrix; Γ is a $m \times k$ matrix; y_t is an $m \times 1$ vector of dependent variables; x_t is a $k \times 1$ vector of explanatory variables; and ϵ_t is an $m \times 1$ vector of disturbance terms. If we take the expectations on both sides, the model to be estimated here is the same as in the GMM estimation represented by (3.1), except that here the functions are linear. Non-linearity may pose a problem in deriving the log-likelihood function for (3.6).

Suppose there are T observations. Then (3.6) can be rewritten as

$$BY' + \Gamma X' = E', \tag{3.7}$$

where Y is $T \times m$; X is $T \times k$, and E is $T \times m$. Assuming normal and serially uncorrelated ϵ_t with the covariance matrix Σ, the concentrated log-likelihood function can be derived (see Chow 1983a, 170–171) as

$$\log L(\psi) = const. + n \log |B| - \frac{n}{2} \log |\Sigma|, \tag{3.8}$$

with the ML estimator of Σ given by

$$\widehat{\Sigma} = n^{-1}(BY' + \Gamma X')(YB' + X\Gamma'). \tag{3.9}$$

The ML estimator of ψ is the one that maximizes log $L(\psi)$ in (3.8). The asymptotic standard deviation of estimated parameters can be inferred from the following variance-covariance matrix of $\widehat{\psi}$ (see Hamilton 1994, 143):

$$E(\widehat{\psi} - \psi)(\widehat{\psi} - \psi)' \cong -\left[\frac{\partial^2 L(\psi)}{\partial \psi \partial \psi'}\right]^{-1}. \tag{3.10}$$

3.4 The Estimation Strategy

In practice, using GMM or ML to estimate a stochastic dynamic model is rather complicated. The first problem that we need to discuss is the restrictions on the estimation.

One proper restriction is that the state equation and a first-order condition are derived either as a Euler equation or from the Lagrangian.[4] Yet, most first-order conditions are extremely complicated and may include some auxiliary variables, such as the Lagrangian multiplier, that are not observable. This seems to suggest that the restrictions for a stochastic dynamic optimization model should typically be represented by the state equation and the control equation derived from the dynamic optimization problem.

The derivation of the control equations in approximate form from a dynamic optimization problem is a complicated process. As discussed in the previous chapters, often a numerical procedure is required. For an approximation method, the linearization of the system at its steady state is needed. The linearization and the derivation of the control equation, possibly through an iterative procedure, often leave it unclear how the parameters to be estimated are related to the model's restrictions and, hence, to the objective function in estimation, such as (3.3) and (3.8). Therefore one usually cannot analytically derive the first-order condition of minimizing (3.3) or maximizing (3.8) with respect to the parameters. Furthermore, using first-order conditions to minimize (3.3) and maximize (3.8) may lead only to a local optimum, which is quite possible, since the system to be estimated is often nonlinear in parameters. Consequently, searching a parameter space becomes the only possible way to find the optimum.

Our search process includes the following recursive steps:

- Step 1. Start with an initial guess of ψ and use an appropriate method of dynamic optimization to derive the decision rules.
- Step 2. Use the state equation and the derived control equation to calculate the value of the objective function.
- Step 3. Apply some optimization algorithm to change the initial guess of ψ and start again with step 1.

Using this strategy to estimate a stochastic dynamic model, one needs to employ an optimization algorithm to search the parameter space recursively. The conventional optimization algorithms,[5] such as Newton–Raphson and related methods, may not serve our purpose well because of the possible existence of multiple local optima. We thus need to employ a global optimization algorithm to execute the estimation process as described above. One possible candidate is simulated annealing, which will be discussed in the next section.

3.5 A Global Optimization Algorithm: Simulated Annealing

The idea of simulated annealing was initially proposed by Metropolis et al. (1953) and later developed by Vanderbilt and Louie (1984), Bohachevsky et al. (1986), and Corana et al. (1987). The algorithm operates through an iterative random search for the optimal variables of an objective function within an appropriate space. It moves uphill and downhill with a varying step size in order to escape local optima. The step size is narrowed so that the random search is confined to an ever smaller region when the global optimum is approached.

The simulated annealing algorithm has been tested by Goffe et al. (1992). For this test, Goffe et al. computed a test function with two optima provided by Judge et al. (1985, 956–957). Comparing it with conventional algorithms, they found that conventional algorithms are 52–60 percent successful in reaching the global optimum, whereas simulated annealing is 100 percent efficient. We thus believe that the algorithm may serve our purpose well.

Let $f(x)$ be a function that is to be maximized, and $x \in S$, where S is the parameter space with the dimensions equal to the number of structural parameters that need to be estimated. The space S should be defined from the economic viewpoint and by computational convenience. The algorithm starts with an initial parameter vector x^0. Its value $f^0 = f(x^0)$ is calculated and recorded. Subsequently, we set the optimum x and $f(x)$—denoted by x_{opt} and f_{opt}, respectively—to x^0 and $f(x^0)$. Other initial conditions include the initial step length (a vector with the same dimension as x), denoted by v^0, and an initial temperature (a scalar), denoted by T^0.

The new variable, x', is chosen by varying the ith element of x^0 such that

$$x'_i = x^0_i + r \cdot v^0_i, \tag{3.11}$$

where r is a uniformly distributed random number in $[-1, 1]$. If x' is not in S, repeat (3.11) until x' is in S. The new function value $f' = f(x')$ is then computed. If f' is larger than f^0, x' is accepted. If not, the Metropolis criterion,[6] denoted as p, is used to decide on acceptance, where

$$p = e^{(f'-f)/T^0}. \tag{3.12}$$

This p is compared with p', a uniformly distributed random number from [0,1]. If p is greater than p', x' is accepted. In addition, f' should be compared with the updated f_{opt}. If it is larger than f_{opt}, both x_{opt} and f_{opt} are replaced by x' and f'.

The above steps, starting with (3.11), should be undertaken and repeated N_S times[7] for each i. Subsequently, the step length is adjusted. The ith element of the new step-length vector (denoted as v'_i) depends on its number of acceptances (denoted as n_i) in the last N_S times of the above repetition, and is given by

$$v'_i = \begin{cases} v^0_i \left[1 + \frac{1}{0.4} c_i \left(n_i/N_S - 0.6\right)\right] & \text{if } n_i > 0.6 N_S; \\ v^0_i \left[1 + \frac{1}{0.4} c_i \left(0.4 - n_i/N_S\right)\right]^{-1} & \text{if } n_i < 0.4 N_S; \\ v^0_i & \text{if } 0.4 N_S \le n_i \le 0.6 N_S, \end{cases} \tag{3.13}$$

where c_i is suggested to be 2 (Corona et al. 1987) for all i. With the new selected step-length vector, one goes back to (3.11) and starts a new round of iteration. After another N_S repetitions, the step length will be readjusted. These adjustments to each v_i should be performed N_T times.[8] We then adjust the temperature. The new temperature (denoted as T') will be

$$T' = R_T T^0, \tag{3.14}$$

with $0 < R_T < 1$.[9] With this new temperature T', we go back again to (3.11). This time, the initial variable x^0 is replaced by the updated x_{opt}. Of course, the temperature will be reduced further after one additional N_T times of adjusting the step length of each i.

For convergence, the step length in (3.11) is required to be very small. In (3.13), whether the new selected step length is enlarged or not depends on the corresponding number of acceptances. The number of acceptances n_i is determined not only by whether the new selected x_i increases the value of objective function, but also by the Metropolis criteria, which in turn depend on the temperature. Thus a convergence will ultimately be achieved with the continuous reduction of the temperature. The algorithm will end by comparing the value of f_{opt} for the last N_ϵ times (suggested to be 4) when readjustment of the temperature is attempted.

3.6 Conclusions

In this chapter, we first introduced the calibration method, which has often been employed in the assessment of a stochastic dynamic model. We then, based on approximation methods, presented an estimation strategy for stochastic dynamic models employing time series data. We introduced both the General Method of Moments (GMM) and the Maximum Likelihood (ML) estimations as strategies to match the dynamic decision model with time series data. Although both strategies permit estimates of the parameters involved, often a global optimization algorithm, such as simulated annealing, must be employed to detect the correct parameters.

3.7 Appendix: A Sketch of the Computer Program for Estimation

The algorithm we describe here is written in GAUSS. The program consists of three parts. The first part regards some necessary steps in the data processing after loading the original data. The second part is the procedure that calculates the value of the objective function for the estimation. The input of this procedure is the structural parameters, and the activation of this procedure generates the value of the objective function. We denote this procedure as OBJF(φ). The third part, which is the main part of this program, is the simulated annealing. Of these three parts, we shall describe only the simulated annealing.

```
  {Set initial conditions for simulated annealing}
DO UNTIL convergence;
  t = t + 1;
  DO N_T times;
    n = 0;                 /*set the vector for recording no. of acceptances*/
    DO N_s times;
      i = 0;
      DO UNTIL i = the dimensions of φ;
        i = i + 1;
        HERE: φ_i = φ_i + rv_i;
```

```
  φ' = {as if current φ except the ith element to be φ'_i};
  IF φ' is not in S;
    GOTO HERE;
    ELSE;
    CONTINUE;
  ENDIF;
  f = OBJF(φ'); /*f is the value of the objective function to be minimized*/
  p = exp[(f' − f)/T];                    /*p is the Metropolis criteria*/
  IF f' > f or p > p'
    φ = φ';
    f = f';
    n_i = n_i + 1;
    ELSE;
    CONTINUE;
  ENDIF;
    IF f' > f_opt;
    φ_opt = φ';
    f_opt = f';
    ELSE;
    CONTINUE;
    ENDIF;
  ENDO;
ENDO;
i = 0;
  {define the new step size, v', according to n_i}
  v = v';
ENDO;
IF change of f_opt < ε in last N_ε times;
  REPORT φ_opt and f_opt;
  BREAK
  ELSE
  T = R_T T;
  CONTINUE;
ENDIF;
ENDO;
```

PART II

THE STANDARD STOCHASTIC DYNAMIC OPTIMIZATION MODEL

4

Real Business Cycles: Theory and Solutions

4.1 Introduction

The real business cycle model as a prototype of a stochastic dynamic macromodel has influenced quantitative macromodeling enormously since the mid-1980s. Its concepts and methods have diffused into mainstream macroeconomics. The criticism of the performance of Keynesian macroeconometric models in the 1970s and the associated rational expectation revolution pioneered by Lucas (1976) initiated this development. The real business cycle analysis now occupies a major position in the curriculum of many graduate programs. To some extent, this approach has become a new orthodoxy of macroeconomics.

The central argument of real business cycle theorists is that economic fluctuations are caused primarily by real factors. Kydland and Prescott (1982) and Long and Plosser (1983) first strikingly illustrated this idea in a simple representative agent optimization model with market clearing, rational expectation, and no monetary factors. Stokey et al. (1989) further illustrate that such a type of model could be viewed as an Arrow–Debreu economy so that the model could be established on a solid microfoundation with many (identical) agents. Therefore, the RBC analysis can also be regarded as a general equilibrium approach to macrodynamics.

This chapter introduces the RBC model by first describing its microeconomic foundation as set out by Stokey et al. (1989). We then present the standard RBC model as formulated in King et al. (1988a, 1988b). A model of this kind will be used repeatedly in the subsequent chapters in various ways. It will then be solved after being parameterized by the standard values of the model's structural parameters.

4.2 The Microfoundation

The standard real business cycle model assumes a representative agent who solves a resource allocation problem over an infinite time horizon via dynamic optimization. It is argued that "the solutions to planning problems of this type can, under appropriate conditions, be interpreted as predictions about the behavior of market economies" (Stokey et al. 1989, 22).

To establish the connection to the competitive equilibrium of the Arrow–Debreu economy,[1] several assumptions should be made for a hypothetical economy. First, the households in the economy are identical, all with the same preference, and firms are also identical, all producing a common output with the same constant returns to scale technology. With this assumption, the resource allocation problem can be viewed as an optimization problem of a representative agent.

Second, as in the Arrow–Debreu economy, the trading process is assumed to be "once-and-for-all."

> Finally, assume that all transactions take place in a single once-and-for-all market that meets in period 0. All trading takes place at that time, so all prices and quantities are determined simultaneously. No further trades are negotiated later. After this market has closed, in periods $t = 0, 1, \ldots, T$, agents simply deliver the quantities of factors and goods they have contracted to sell and receive those they have contracted to buy. (Stokey et al. 1989, 23)

The third assumption regards ownership. It is assumed that the household owns all factors of production and all shares of the firm. Therefore, in each period the household sells factor services to the firm. The revenue from selling factors can be used only to buy the goods produced by the firm, either for consuming or for accumulating as capital. The representative firm owns nothing. In each period it simply hires capital and labor on a rental basis to produce output, sells the output, and transfers any profit back to the household.

4.2.1 The Decision of the Household

At the beginning of period 0 when the market is open, the household is given the price sequence $\{p_t, w_t, r_t\}_{t=0}^{\infty}$ at which he (or she) will choose the sequence of output demand and input supply $\{c_t^d, i_t^d, n_t^s, k_t^s\}_{t=0}^{\infty}$ that maximizes the discounted utility:

$$\max E_0 \left[\sum_{t=0}^{\infty} \beta^t U\left(c_t^d, n_t^s\right) \right], \tag{4.1}$$

subject to

$$p_t\left(c_t^d + i_t^d\right) = p_t\left(r_t k_t^s + w_t n_t^s\right) + \pi_t \tag{4.2}$$

$$k_{t+1}^s = (1-\delta)k_t^s + i_t^d. \tag{4.3}$$

Above, δ is the depreciation rate; β is the discounted factor; π_t is the expected dividend; c_t^d and i_t^d are the demands for consumption and investment, respectively, and n_t^s and k_t^s are the supplies of labor and capital stock, respectively. Note that (4.2) can be regarded as a budget constraint. The equality holds because of the assumption $U_c > 0$. Next, we consider how the representative household calculates π_t. It is reasonable to assume that

$$\pi_t = p_t\left(\widehat{y}_t - w_t\widehat{n}_t - r_t\widehat{k}_t\right), \tag{4.4}$$

where $\widehat{y}_t$, $\widehat{n}_t$, and $\widehat{k}_t$ are the realized output, labor, and capital expected by the household at given price sequence $\left\{p_t, w_t, r_t\right\}_{t=0}^{\infty}$. Thus, assuming that the household knows the production function while expecting that the market will be cleared at the given price sequence $\left\{p_t, w_t, r_t\right\}_{t=0}^{\infty}$, (4.4) can be rewritten as

$$\pi_t = p_t\left[f\left(k_t^s, n_t^s, \hat{A}_t\right) - w_t n_t^s - r_t k_t^s\right]. \tag{4.5}$$

Above, $f(\cdot)$ is the production function and $\hat{A}_t$ is the expected technology shock. Explaining π_t in (4.2) in terms of (4.5) and then substituting from (4.3) to eliminate i_t^d, we obtain

$$k_{t+1}^s = (1-\delta)k_t^s + f\left(k_t^s, n_t^s, \hat{A}_t\right) - c_t^d. \tag{4.6}$$

Note that (4.1) and (4.6) represent the standard RBC model, although it specifies only one side of the market: output demand and input supply. Given the initial capital stock k_0^s, the solution of this model is the sequence of plans $\left\{c_t^d, i_t^d, n_t^s, k_{t+1}^s\right\}_{t=0}^{\infty}$, where k_t^s is implied by (4.6) and

$$c_t^d = G_c\left(k_t^s, \hat{A}_t\right) \tag{4.7}$$

$$n_t^s = G_n\left(k_t^s, \hat{A}_t\right) \tag{4.8}$$

$$i_t^d = f\left(k_t^s, n_t^s, \hat{A}_t\right) - c_t^d. \tag{4.9}$$

4.2.2 The Decision of the Firm

Given the same price sequence $\left\{p_t, w_t, r_t\right\}_{t=0}^{\infty}$, and also the sequence of expected technology shocks $\{\hat{A}_t\}_{t=0}^{\infty}$, the problem faced by the representative firm is to choose input demand and output supply $\{y_t^s, n_t^d, k_t^d\}_{t=0}^{\infty}$. However, since the firm simply rents capital and hires labor on a period-by-period basis, its optimization problem is equivalent to a series of one-period maximization (Stokey et al. 1989, 25):

$$\max \ p_t \left(y_t^s - r_t k_t^d - w_t n_t^d\right),$$

subject to

$$y_t^s = f\left(k_t^d, n_t^d, \hat{A}_t\right), \tag{4.10}$$

where $t = 0, 1, 2, \ldots, \infty$. The solution to this optimization problem satisfies

$$r_t = f_k\left(k_t^d, n_t^d, \hat{A}_t\right)$$

$$w_t = f_n\left(k_t^d, n_t^d, \hat{A}_t\right).$$

This first-order condition allows us to derive the equations of input demand k_t^d and n_t^d:

$$k_t^d = k\left(r_t, w_t, \hat{A}_t\right) \tag{4.11}$$

$$n_t^d = n\left(r_t, w_t, \hat{A}_t\right). \tag{4.12}$$

4.2.3 The Competitive Equilibrium and the Walrasian Auctioneer

A competitive equilibrium can be described as a sequence of prices $\{p_t^*, w_t^*, r_t^*\}_{t=0}^{\infty}$ at which the two market forces (demand and supply) are equalized in all three markets:

$$k_t^d = k_t^s \tag{4.13}$$

$$n_t^d = n_t^s \tag{4.14}$$

$$c_t^d + i_t^d = y_t^s \tag{4.15}$$

for all t's, $t = 0, 1, 2, \ldots, \infty$. The economy is at the competitive equilibrium if

$$\{p_t, w_t, r_t\}_{t=0}^{\infty} = \{p_t^*, w_t^*, r_t^*\}_{t=0}^{\infty}.$$

Using equations (4.6)–(4.12), one can easily prove the existence of $\{p_t^*, w_t^*, r_t^*\}_{t=0}^{\infty}$ that satisfies the equilibrium condition (4.13)–(4.15).

The RBC literature usually does not explain how the equilibrium is achieved. Implicitly, it is assumed that there exists an auctioneer in the market who adjusts the price toward the equilibrium. This process—often called the *tâtonnement* process in Walrasian economics—is a common solution to the adjustment problem within the neoclassical general equilibrium framework.

4.2.4 The Contingency Plan

It is not difficult to determine that the sequence of equilibrium prices $\{p_t^*, w_t^*, r_t^*\}_{t=0}^{\infty}$ depends on the expected technology shock $\{\hat{A}_t\}_{t=0}^{\infty}$. This creates a problem of how to express the equilibrium prices and the equilibrium demand and supply that are supposed to prevail at the beginning of period 0 when the technology shocks from period 1 on are unobserved. The Real Business Cycle theorists circumvent this problem skillfully and ingeniously. Their approach is to use the "contingency plan."

> In the stochastic case, however, this is not a sequence of numbers but a sequence of contingency plans, one for each period. Specifically, consumption c_t, and end-of-period capital k_{t+1} in each period $t = 1, 2, \ldots$ are contingent on the realization of the shocks $z_1, z_2, \ldots, z_t$. This sequence of realization is information that is available when the decision is being carried out but is unknown in period 0 when the decision is being made. Technically, then, the planner chooses among sequence of functions. (Stokey et al., 1989, 17)

Thus the sequence of equilibrium prices and the sequence of equilibrium demand and supply are contingent on the realization of the shock even though the corresponding decisions are made at the beginning of period 0.

4.2.5 The Dynamics

Assuming that all the decisions are contingent on the future shock $\{A_t\}_{t=0}^{\infty}$, and all the prices are at their equilibrium, the dynamics of our hypothetical economy can be fully described by the following equations regarding the realized consumption, employment, output, investment, and capital stock:

$$c_t = G_c(k_t, A_t) \tag{4.16}$$

$$n_t = G_n(k_t, A_t) \tag{4.17}$$

$$y_t = f(k_t, n_t, A_t) \tag{4.18}$$

$$i_t = y_t - c_t \tag{4.19}$$

$$k_{t+1} = (1-\delta)k_t + f(k_t, n_t, A_t) - c_t, \tag{4.20}$$

given the initial condition k_0 and the sequence of technology shock $\{A_t\}_{t=0}^{\infty}$. This provides another important property of the RBC economy. Although the model specifies the decision behaviors for both the household and the firm, and therefore the demand and supply, for all major three markets—output, capital, and labor—the dynamics of the economy is reflected only by the household behavior, which concerns only one side of the market forces: output demand and input supply. The decision of the firm does not have any impact! This is certainly due to the equilibrium feature of the model specification.

4.3 The Standard RBC Model

4.3.1 The Model Structure

The hypothetical economy we presented in section 4.2 is only for explaining the theory (from a microeconomic point of view) behind the standard RBC economy. The model specified in (4.16)–(4.20) is not testable with empirical data, not only because we do not specify the stochastic process of $\{A_t\}_{t=0}^{\infty}$ but also because we do not introduce the growth factor. For an empirically testable standard RBC model, we employ here the specifications of a model formulated by King et al. (1988a, 1988b). This empirically oriented formulation will be used repeatedly in the subequent chapters of this volume.

Let K_t denote the aggregate capital stock, Y_t aggregate output, and C_t aggregate consumption. The capital stock in the economy follows the transition law:

$$K_{t+1} = (1 - \delta)K_t + Y_t - C_t, \tag{4.21}$$

where δ is the depreciation rate. Assume that the aggregate production function takes the form

$$Y_t = A_t K_t^{1-\alpha}(N_t X_t)^{\alpha}, \tag{4.22}$$

where N_t is per capita working hours; α is the share of labor in the production function; A_t is the temporary shock in technology; and X_t is the permanent shock that follows a growth rate γ. Here, X_t includes not only the growth in labor force but also the growth in productivity. Apparently the model is nonstationary due to X_t. To transform the model into a stationary formulation, we divide both sides of equation (4.21) by X_t (when Y_t is expressed by (4.22)):

$$k_{t+1} = \frac{1}{1+\gamma}\left[(1-\delta)k_t + A_t k_t^{1-\alpha}(n_t \bar{N}/0.3)^{\alpha} - c_t\right], \tag{4.23}$$

where by definition, $k_t \equiv K_t/X_t$, $c_t \equiv C_t/X_t$ and $n_t \equiv 0.3N_t/\bar{N}$, with $\bar{N}$ the sample mean of N_t. Note that n_t is often regarded as the normalized hours.

The sample mean of n_t is equal to 30%, which, as pointed out by Hansen (1985), is the average percentage of hours attributed to work.

The representative agent in the economy is assumed to make the decision sequence $\{c_t\}_{t=0}^{\infty}$ and $\{n_t\}_{t=0}^{\infty}$ so that

$$\max\ E_0 \sum_{t=0}^{\infty} \beta^t \left[\log c_t + \theta \log\,(1 - n_t)\right], \tag{4.24}$$

subject to the state equation (4.23). The exogenous variable in this model is the temporary shock A_t, which may follow an AR(1) process:

$$A_{t+1} = a_0 + a_1 A_t + \epsilon_{t+1}, \tag{4.25}$$

with ϵ_t to be an i.i.d. innovation.

There is no possibility of deriving the exact solution with the standard recursive method. We therefore have to rely on an approximate solution. For this, we first derive the first-order conditions.

4.3.2 The First-Order Conditions

As we discussed in the previous chapters, there are two types of first-order conditions, the Euler equation and the equation from the Lagrangian. The Euler equation is not used in our suggested solution method. We nevertheless present it here as an exercise and demonstrate that the two first-order conditions are virtually equivalent.

The Euler Equation

To derive the Euler equation, our first task is to transform the model into a setting in which the state variable k_t does not appear in $F(\cdot)$, as we discussed in chapters 1 and 2. This can be done by assuming k_{t+1} (instead of c_t), along with n_t, as model's decision variables. In this case, the objective function takes the form

$$\max\ E_0 \sum_{t=0}^{\infty} \beta^t U(k_{t+1}, n_t, k_t, A_t),$$

where

$$\begin{aligned} U(k_{t+1}, n_t, k_t, A_t) = {} & \log\left[(1-\delta)k_t + y_t - (1+\gamma)k_{t+1}\right] \\ & + \theta \log\,(1 - n_t). \end{aligned} \tag{4.26}$$

Here we have used (4.23) to express c_t in the utility function; y_t is the stationary output via the equation

$$y_t = A_t k_t^{1-\alpha} (n_t \bar{N} / 0.3)^{\alpha}. \tag{4.27}$$

Given such an objective function, the state equation (4.23) can simply be ignored in deriving the first-order condition. The Bellman equation in this case can be written as

$$V(k_t, A_t) = \max_{k_{t+1}, n_t} U(k_{t+1}, n_t, k_t, A_t) + \beta E\left[V(k_{t+1}, A_{t+1})\right]. \tag{4.28}$$

The necessary condition for maximizing the right side of the Bellman equation (4.28) is given by

$$\frac{\partial U}{\partial k_{t+1}}(k_{t+1}, n_t, k_t, A_t) + \beta E\left[\frac{\partial V}{\partial k_{t+1}}(k_{t+1}, A_{t+1})\right] = 0; \tag{4.29}$$

$$\frac{\partial U}{\partial n_t}(k_{t+1}, n_t, k_t, A_t) = 0. \tag{4.30}$$

Meanwhile, the application of the Benveniste–Scheinkman formula gives

$$\frac{\partial V}{\partial k_t}(k_t, A_t) = \frac{\partial U}{\partial k_t}(k_{t+1}, n_t, k_t, A_t). \tag{4.31}$$

Using (4.31) to express $\frac{\partial V}{\partial k_{t+1}}(k_{t+1}, A_{t+1})$ in (4.29), we obtain

$$\frac{\partial U}{\partial k_{t+1}}(k_{t+1}, n_t, k_t, A_t) + \beta E\left[\frac{\partial U}{\partial k_{t+1}}(k_{t+2}, n_{t+1}, k_{t+1}, A_{t+1})\right] = 0. \tag{4.32}$$

From equations (4.23) and (4.26),

$$\frac{\partial U}{\partial k_{t+1}}(k_{t+1}, n_t, k_t, A_t) = -\frac{1+\gamma}{c_t};$$

$$\frac{\partial U}{\partial k_{t+1}}(k_{t+2}, n_{t+1}, k_{t+1}, A_{t+1}) = \frac{(1-\delta)k_{t+1} + (1-\alpha)y_{t+1}}{k_{t+1}c_{t+1}};$$

$$\frac{\partial U}{\partial n_t}(k_{t+1}, n_t, k_t, A_t) = \frac{\alpha y_t}{n_t c_t} - \frac{\theta}{1-n_t}.$$

Substituting the above expressions into (4.32) and (4.30), we establish the following Euler equations:

$$-\frac{1+\gamma}{c_t} + \beta E\left[\frac{(1-\delta)k_{t+1} + (1-\alpha)y_{t+1}}{k_{t+1}c_{t+1}}\right] = 0; \tag{4.33}$$

$$\frac{\alpha y_t}{n_t c_t} - \frac{\theta}{1-n_t} = 0. \tag{4.34}$$

The First-Order Condition Derived from the Lagrangian

Next, we turn to the first-order condition derived from the Lagrangian. Define the Lagrangian:

$$L = \sum_{t=0}^{\infty} \beta^t \left[\log (c_t) + \theta \log (1 - n_t)\right]$$

$$- \sum_{t=0}^{\infty} E_t \left\{ \beta^{t+1} \lambda_{t+1} \left[k_{t+1} - \frac{1}{1+\gamma} \left((1-\delta) k_t + A_t k_t^{1-\alpha} (n_t \bar{N} / 0.3)^{\alpha} - c_t \right) \right] \right\}.$$

Setting to 0 the derivatives of L with respect to c_t, n_t, k_t, and λ_t, one obtains the following first-order condition:

$$\frac{1}{c_t} - \frac{\beta}{1+\gamma} E_t \lambda_{t+1} = 0; \tag{4.35}$$

$$\frac{-\theta}{1-n_t} + \frac{\alpha \beta y_t}{(1+\gamma) n_t} E_t \lambda_{t+1} = 0; \tag{4.36}$$

$$\frac{\beta}{1+\gamma} E_t \lambda_{t+1} \left[(1-\delta) + \frac{(1-\alpha) y_t}{k_t} \right] = \lambda_t; \tag{4.37}$$

$$k_{t+1} = \frac{1}{1+\gamma} \left[(1-\delta) k_t + y_t - c_t \right], \tag{4.38}$$

with y_t again given by (4.27).

Next we try to demonstrate that the first-order conditions (4.33)–(4.34) and (4.35)–(4.38) are virtually equivalent. This can be done as follows. First, expressing $[\beta/(1+\gamma)] E_t \lambda_{t+1}$ in terms of $1/c_t$ (which is implied by (4.35)), from (4.37) we obtain

$$\lambda_t = \frac{(1-\delta) k_t + (1-\alpha) y_t}{k_t c_t}.$$

This further indicates that

$$E_t \lambda_{t+1} = \frac{(1-\delta) k_{t+1} + (1-\alpha) y_{t+1}}{k_{t+1} c_{t+1}}. \tag{4.39}$$

Substituting (4.39) into (4.35), we obtain the first Euler equation (4.33). Second, expressing $[\beta/(1+\gamma)] E_t \lambda_{t+1}$ again in terms of $1/c_t$, and substituting it into (4.36), we verify the second Euler equation (4.34).

4.3.3 The Steady States

Next we try to derive the corresponding steady states. The steady state A_t is simply determined from (4.25). The other steady states are given by the following proposition.

Proposition 4 *Assume A_t has a steady state $\bar{A}$. Equations (4.35)–(4.38), along with (4.27), when evaluated in terms of their certainty equivalence forms,* *determine at least two steady states: one is boundary, denoted as* $(\bar{c}_b, \bar{n}_b, \bar{k}_b, \bar{y}_b, \bar{\lambda}_b)$*, and the other is interior, denoted as* $(\bar{c}_i, \bar{n}_i, \bar{k}_i, \bar{y}_i, \bar{\lambda}_i)$*. In particular,*

$$\bar{c}_b = 0,$$
$$\bar{n}_b = 1,$$
$$\bar{\lambda}_b = \infty,$$
$$\bar{k}_b = \left[\bar{A}/(\delta+\gamma)\right]^{1/\alpha}\left(\bar{N}/0.3\right),$$
$$\bar{y}_b = (\delta+\gamma)\bar{k}_b;$$

and

$$\bar{n}_i = \alpha\phi/\left[(\alpha+\theta)\phi - (\delta+\gamma)\theta\right],$$
$$\bar{k}_i = \bar{A}^{1/\alpha}\phi^{-1/\alpha}\bar{n}_i\left(\bar{N}/0.3\right)_i$$
$$\bar{c}_i = (\phi-\delta-\gamma)\bar{k}_i,$$
$$\bar{\lambda}_i = (1+\gamma)/\beta\bar{c}_i$$
$$\bar{y}_i = \phi\bar{k}_i,$$

where

$$\phi = \frac{(1+\gamma)-\beta(1-\delta)}{\beta(1-\alpha)}. \tag{4.40}$$

We have used the first-order condition from the Lagrangian to derive the above two steady states. Since the two first-order conditions are virtually equivalent, we expect that the same steady states can be derived from the Euler equations.[2]

4.4 Solving the Standard Model with Standard Parameters

To obtain the solution path of the standard RBC model, we first specify the values of the structural parameters defined in the model. These are displayed in table 4.1. These parameters are close to the standard parameters, which can often be found in the RBC literature.[3] More detailed discussion regarding parameter selection and estimation will be provided in chapter 5.

Table 4.1. Parameterizing the Standard RBC Model

α	γ	β	δ	θ	$\bar{N}$	a_0	a_1	σ_ϵ
0.58	0.0045	0.9884	0.025	2	480	0.0333	0.9811	0.0189

The solution method that we employ is the method of linear-quadratic approximation with our suggested algorithm, as discussed in chapter 1. Assume that the decision rule take the form

$$c_t = G_{11}A_t + G_{12}k_t + g_1 \tag{4.41}$$

$$n_t = G_{21}A_t + G_{22}k_t + g_2. \tag{4.42}$$

Our first step is to compute the first- and second-order partial derivatives with respect to F and U, where

$$F(k, c, n, A) = \frac{1}{1+\gamma}\left[(1-\delta)k + Ak^{1-\alpha}(n\bar{N}/0.3)^{\alpha} - c\right]$$

$$U(c, n) = \log(c) + \theta \log(1-n).$$

All these partial derivatives, along with the steady states, can be used as the inputs in the GAUSS procedure provided in appendix II of chapter 1. Executing this procedure will allow us to compute the undetermined coefficients G_{ij} and $g_i (i, j = 1, 2)$ in the decision rule, as expressed in (4.41) and (4.42). In figures 4.1 and 4.2, we illustrate the solution paths, one for the

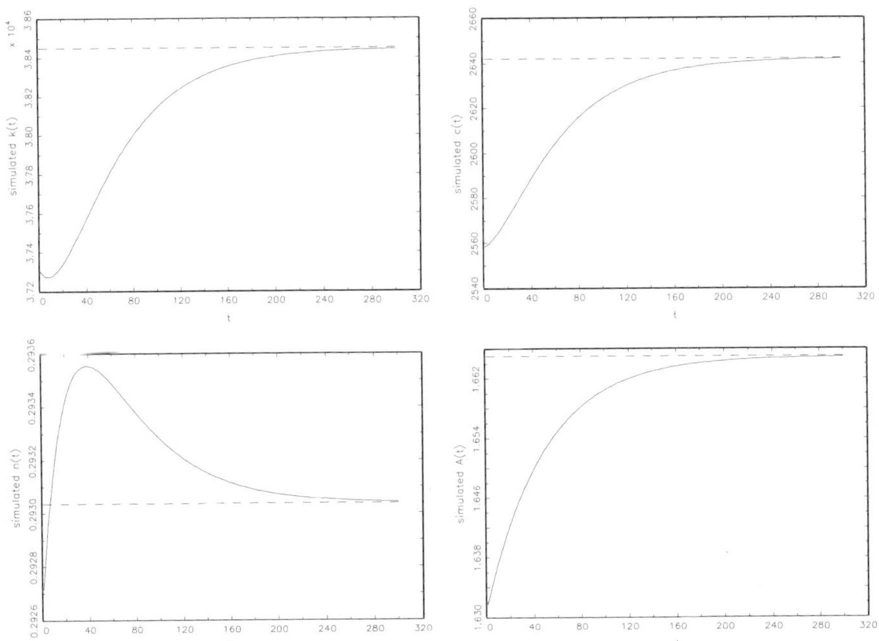

Figure 4.1. The Deterministic Solution to the Benchmark RBC Model for the Standard Parameters

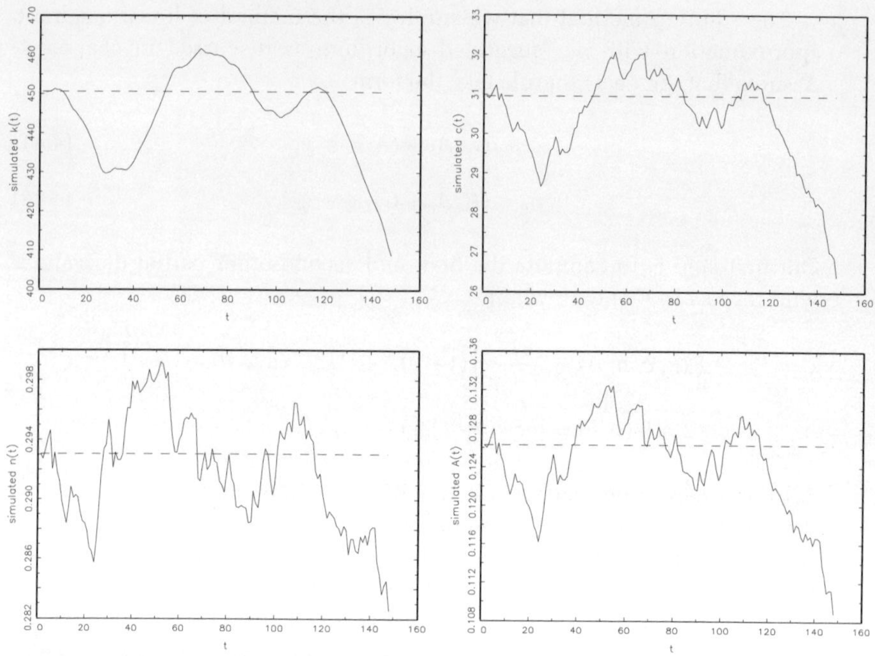

Figure 4.2. The Stochastic Solution to the Benchmark RBC Model for the Standard Parameters

deterministic case and the other for the stochastic case, for the variables k_t, c_t, n_t, and A_t.

Elsewhere (see Gong and Semmler 2001), we have compared these solution paths with those computed by Campbell's (1994) log-linear approximate solution. The two solutions are so close that one can hardly observe the differences.

4.5 The Generalized RBC Model

In recent work, stochastic dynamic optimization models of the general equilibrium type that go beyond the standard model (as discussed in section 4.3) have been presented in the literature. The recent models are more demanding in terms of solution and estimation methods. Although we will not attempt to estimate those more generalized versions, it is worth presenting the main structure of the generalized models and demonstrating how they can be solved by using dynamic programming as introduced in chapter 1, section 1.6. Since those generalized versions can easily give rise to multiple equilibria and history dependence, we will return to them in chapter 7.

4.5.1 The Model Structure

The generalization of the standard RBC model is usually undertaken with respect to preferences or with respect to the technology. With respect to preferences, utility functions such as

$$U(C, N) = \frac{\left[C\, exp\left(\frac{-N^{1+\chi}}{1+\chi}\right)\right]^{1-\sigma} - 1}{1-\sigma} \tag{4.43}$$

are used.[4] The utility function (4.43) with consumption C and labor effort N as arguments is nonseparable in consumption and leisure. From (4.43) we can obtain a separable utility function such as

$$U(C, N) = \frac{C^{1-\sigma}}{1-\sigma} - \frac{N^{1+\chi}}{1+\chi}, \tag{4.44}$$

which is additively separable in consumption and leisure.[5]

Moreover, by setting $\sigma = 1$ we obtain simplified preferences in log-utility and leisure:

$$U(C, N) = \log C - \frac{N^{1+\chi}}{1+\chi}. \tag{4.45}$$

As for production technology and markets, usually the following generalizations are introduced.[6] First, we can allow for increasing returns to scale. Although individual-level private technology generates constant returns to scale with

$$Y_i = AK_i^a L_i^b, \quad a + b = 1,$$

externalities of the form

$$A = (K^a N^b)^{\xi}, \quad \xi \geq 0$$

may allow for an aggregate production function with increasing returns to scale with

$$Y = K^{\alpha} N^{\beta} \quad \alpha > 0, \beta > 0, \alpha + \beta \geq 1, \tag{4.46}$$

whereby $\alpha = (1 + \xi)a$, $\beta = (1 + \xi)b$, and Y, K, N represent total output, the aggregate stock of capital, and labor hours, respectively. The increasing returns to scale technology represented by (4.46) can also be interpreted as a monopolistic competition economy where there are rents arising from inverse demand curves for monopolistic firms.[7]

Another generalization concerning production technology can be undertaken by introducing adjustment cost of investment.[8] We may write

$$\frac{\dot{K}}{K} = \varphi\left(\frac{I}{K}\right) \tag{4.47}$$

with the assumption of

$$\varphi(\delta) = 0, \quad \varphi'(\delta) = 1, \quad \varphi''(\delta) \leq 0. \tag{4.48}$$

Here δ is the depreciation rate of the capital stock. A functional form that satisfies the three conditions of (4.47) is

$$\delta\left[\left(\frac{I}{\delta K}\right)^{1-\varphi} - 1\right]/(1-\varphi). \tag{4.49}$$

For $\varphi = 0$, one has the standard model without adjustment cost,

$$\dot{K} = I - \delta K.$$

The above describes a type of a generalized model that we want to solve.

4.5.2 Solving the Generalized RBC Model

We write the model in continuous time and in its deterministic form:

$$V = \max_{\{C,N\}} \int_{t=0}^{\infty} e^{-\rho t} U(C_t, N_t) dt \tag{4.50}$$

s.t.

$$\frac{\dot{K}}{K} = \varphi\left(\frac{I_t}{K_t}\right), \tag{4.51}$$

where preferences $U(C_t, N_t)$ are chosen as represented by (4.45) and the technology as specified in (4.46) and (4.47)–(4.49). The latter are used in (4.51).

For solving the model with the dynamic programming algorithm as presented in chapter 1, section 1.6, we use the parameters shown in table 4.2.

In order to stay as close as possible to the standard RBC model, we avoid externalities, and therefore presume $\xi = 0$. Preferences and technology (by

Table 4.2. Parameterizing the General Model

a	b	χ	ρ	δ	ξ	φ
0.3	0.7	0.3	0.05	0.1	0	0.05

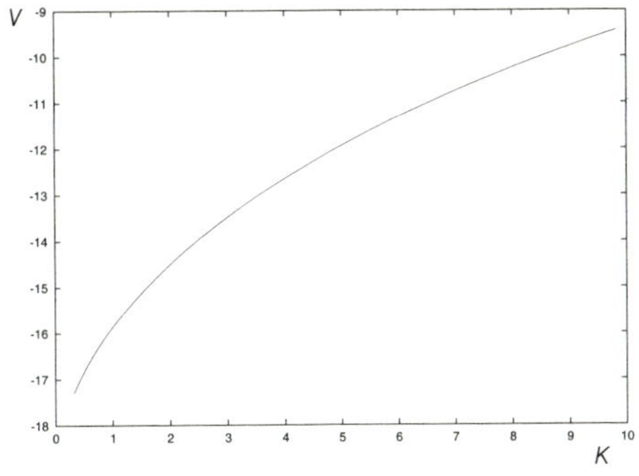

Figure 4.3. Value Function for the General Model

using adjustment cost of capital) take on, however, a more general form. The dynamic decision problem (4.50)–(4.51) is written in continuous time. Its discretization for the use of dynamic programming is undertaken through the Euler procedure. Using the deterministic variant of our dynamic programming algorithm of chapter 1, section 1.6, and a grid for the capital stock, K, in the interval [0.1, 10], we obtain the following value function, representing the total utility along the optimal paths of C and N. (See figure 4.3.)

The out-of-steady-state condition of the two choice variables, consumption and labor effort (depending on feedback from on the state variable capital stock, K), are shown in figure 4.4.

In figure 4.3, the value function is concave. Moreover, in figure 4.4, consumption is low and labor effort is high when capital stock is low (so that capital stock can be built up), and consumption is high and labor effort is low when capital stock is high (so that capital stock will shrink). The dynamics generated by the response of consumption and labor effort to the state variable capital stock, K, will thus lead to a convergence toward an interior steady state of capital stock, consumption, and labor effort.

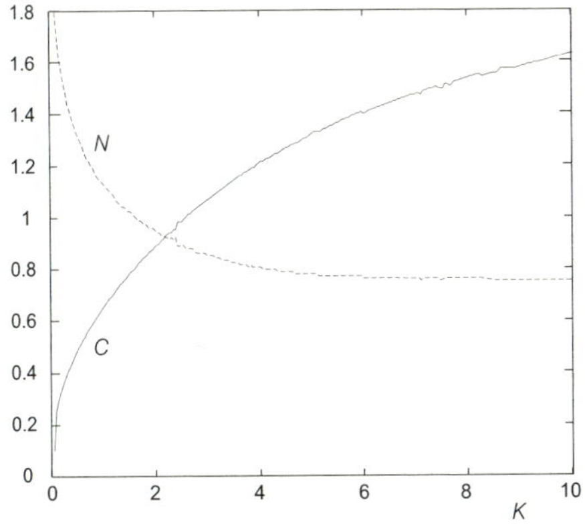

Figure 4.4. Paths of the Choice Variables C and N (depending on K)

4.6 Conclusions

In this chapter we first introduced the intertemporal general equilibrium model on which the standard RBC model is constructed. We sought to reveal the basis of some intertemporal decision problems behind the RBC model. This will be important later in the book. We then introduced the empirically oriented standard RBC model and, based on our previous chapters, presented the solution to the model. This provided the groundwork for an empirical assessment of the RBC model, the task that will be addressed in chapter 5. We also presented a generalized RBC model and solved it by using dynamic programming.

4.7 Appendix: The Proof of Proposition 4

Evaluating (4.35)–(4.38) along with (4.27) in their certainty equivalence form, and assuming all the variables to be at their steady states, we obtain

$$\frac{1}{\bar{c}} - \frac{\beta}{1+\gamma}\bar{\lambda} = 0 \tag{4.52}$$

$$\frac{-\theta}{1-\bar{n}} + \beta\bar{\lambda}\frac{\alpha\bar{y}}{(1+\gamma)\bar{n}} = 0 \tag{4.53}$$

$$\frac{\beta}{1+\gamma}\bar{\lambda}\left[(1-\delta) + \frac{(1-\alpha)\bar{y}}{\bar{k}}\right] = \bar{\lambda} \tag{4.54}$$

$$\bar{k} = \frac{1}{1+\gamma}\left[(1-\delta)\bar{k} + \bar{y} - \bar{c}\right], \tag{4.55}$$

where from (4.27) $\bar{y}$ is given by

$$\bar{y} = \bar{A}\bar{k}^{1-\alpha}\left(\frac{\bar{n}\bar{N}}{0.3}\right)^{\alpha}. \tag{4.56}$$

The derivation of the boundary steady state is trival. Replace $\bar{c}$, $\bar{n}$, and $\bar{\lambda}$ with $\bar{c}_b$, $\bar{n}_b$ and $\bar{\lambda}_b$. Then equations (4.52)–(4.54) are satisfied. Further, $\bar{k}_b$ and $\bar{y}_b$ can be derived from (4.55) and (4.56), given $\bar{c}_b$, $\bar{n}_b$, and $\bar{\lambda}_b$.

Next we try to derive the interior steady state. For notational convenience, we ignore the subscript i, and therefore all the steady-state values are understood as the interior steady states. Let $\bar{y}/\bar{k} = \phi$. By (4.54), we obtain $\bar{y} = \phi\bar{k}$, where ϕ is defined by (4.40) in the proposition. According to (4.56),

$$\frac{\bar{y}}{\bar{n}} = \bar{A}\left(\frac{\bar{y}}{\phi}\right)^{1-\alpha}\left(\frac{\bar{n}\bar{N}}{0.3}\right)^{\alpha}\bar{n}^{-1}$$

$$= \bar{A}\left(\frac{\bar{y}}{\bar{n}}\right)^{1-\alpha}\phi^{\alpha-1}\left(\frac{\bar{N}}{0.3}\right)^{\alpha}$$

$$= \bar{A}^{\frac{1}{\alpha}}\phi^{1-\frac{1}{\alpha}}\left(\frac{\bar{N}}{0.3}\right) \tag{4.57}$$

Therefore,

$$\frac{\bar{k}}{\bar{n}} = \frac{\bar{y}/\bar{n}}{\bar{y}/\bar{k}}$$

$$= \bar{A}^{\frac{1}{\alpha}}\phi^{-\frac{1}{\alpha}}\left(\frac{\bar{N}}{0.3}\right). \tag{4.58}$$

Expressing $\bar{y}$ in terms of $\phi\bar{k}$ and then expressing $\bar{k}$ in terms of (4.58), we thus obtain from (4.55)

$$\bar{c} = (\phi - \delta - \gamma)\bar{k}$$

$$= (\phi - \delta - \gamma)\bar{A}^{\frac{1}{\alpha}}\phi^{-\frac{1}{\alpha}}\left(\frac{\bar{N}}{0.3}\right)\bar{n}. \tag{4.59}$$

Meanwhile, (4.52) and (4.53) imply that

$$\bar{c} = \left(\frac{\alpha}{\theta}\right)(1 - \bar{n})\left(\frac{\bar{y}}{\bar{n}}\right)$$

$$= \left(\frac{\alpha}{\theta}\right)(1 - \bar{n})\bar{A}^{\frac{1}{\alpha}}\phi^{1-\frac{1}{\alpha}}\left(\frac{\bar{N}}{0.3}\right). \tag{4.60}$$

Equating (4.59) and (4.60), we thus solve the steady state of the labor effort $\bar{n}$. Once $\bar{n}$ is solved, $\bar{k}$ can be obtained from (4.58), $\bar{y}$ from (4.57), and $\bar{c}$ from either (4.59) or (4.60). Finally, $\bar{\lambda}$ can be derived from (4.52).

5

Empirics of the Standard Real Business Cycle Model

5.1 Introduction

Many RBC theorists believe that the RBC model is empirically powerful in explaining the stylized facts of business cycles. Moreover, some theorists suggest that even a simple RBC model, such as the standard model presented in chapter 4, despite its rather simple structure, can generate the time series to match the macroeconomic moments from empirically observed time series data. As Plosser pointed out, "The whole idea that such a simple model with no government, no market failures of any kind, rational expectations, no adjustment cost could replicate actual experience this well is very surprising" (King et al. 1999, 961). However, these early assessments have become the subject of various criticisms. In this chapter, we provide a comprehensive empirical assessment of the standard RBC model. Our previous discussion, especially in the first three chapters, has provided a technical preparation for this assessment. We first estimate the standard RBC model and then evaluate the calibration results that have been stated by early RBC theorists. Before we commence our formal study, we will demonstrate the efficiency of our estimation strategy as discussed in chapter 3.

5.2 Estimation with Simulated Data

In this section, we first apply our estimation strategy, using simulated data. The simulated data are shown in figure 4.2, which is generated from a stochastic simulation of our standard model for the given standard parameters reported in table 4.1. The purpose of this estimation is to test whether our suggested estimation strategy works well. If the strategy works well, we expect that the estimated parameters will be close to the standard parameters that we know in advance.

5.2.1 The Estimation Restriction

The standard model implies certain restrictions on the estimation. For the Generalized Method of Moment (GMM) estimation, the model implies the following moment restrictions:[1]

$$E\left[(1+\gamma)k_{t+1} - (1-\delta)k_t - y_t + c_t\right] = 0; \tag{5.1}$$

$$E\left[c_t - G_{11}k_t - G_{12}A_t - g_{13}\right] = 0; \tag{5.2}$$

$$E\left[n_t - G_{21}k_t - G_{22}A_t - g_{23}\right] = 0; \tag{5.3}$$

$$E\left[y_t - A_t k_t^{1-\alpha}(n_t \bar{N}/0.3)^{\alpha}\right] = 0. \tag{5.4}$$

The moment restrictions can be nonlinear, as in (5.4). Yet for the Maximum Likelihood (ML) estimation, the restriction $Bz_t + \Gamma x_t = \varepsilon_t$ (see equation (3.6))[2] must be linear. Therefore, we first linearize (4.27) using the Taylor approximation. This gives us

$$y_t = \frac{\bar{y}}{\bar{A}}A_t + (1-\alpha)\frac{\bar{y}}{\bar{k}}k_t + \alpha\frac{\bar{y}}{\bar{n}}n_t - \bar{y}.$$

We thus obtain for the ML estimation:

$$z_t = \begin{bmatrix} k_t \\ c_t \\ n_t \\ y_t \end{bmatrix}, \quad B = \begin{bmatrix} 1+\gamma & 0 & 0 & 0 \\ -G_{12} & 1 & 0 & 0 \\ -G_{22} & 0 & 1 & 0 \\ -(1-\alpha)\frac{\bar{y}}{\bar{k}} & 0 & -\alpha\frac{\bar{y}}{\bar{n}} & 1 \end{bmatrix}$$

$$x_t = \begin{bmatrix} k_{t-1} \\ c_{t-1} \\ y_{t-1} \\ A_t \\ 1 \end{bmatrix}, \quad \Gamma = \begin{bmatrix} -(1-\delta) & 1 & -1 & 0 & 0 \\ 0 & 0 & 0 & -G_{11} & -g_{13} \\ 0 & 0 & 0 & -G_{21} & -g_{23} \\ 0 & 0 & 0 & -\frac{\bar{y}}{\bar{A}} & \bar{y} \end{bmatrix}$$

5.2.2 Estimation with Simulated Data

Although the model involves many parameters, we will estimate only the parameters α, β, δ, and θ. These are the parameters that are empirically

Table 5.1. GMM and ML Estimations Using Simulated Data

	α	β	δ	θ
True	0.58	0.9884	0.025	2
ML Estimation	0.5781 (2.4373E−006)	0.9946 (5.9290E−006)	0.0253 (3.4956E−007)	2.1826 (3.7174E−006)
1st Step GMM	0.5796 (6.5779E−005)	0.9821 (0.00112798)	0.02500 (9.6093E−006)	2.2919 (0.006181723)
2nd Step GMM	0.5800 (2.9958E−008)	0.9884 (3.4412E−006)	0.02505 (9.7217E−007)	2.000 (4.6369E−006)

unknown and thus need to be estimated when we later turn to the estimation using empirical data.[3] Table 5.1 reports our estimations with the standard deviation included in parentheses.

The estimations from both methods are quite satisfying. All estimated parameters are close to the true parameters, the parameters that we used to generate the data. This demonstrates the efficiency of our estimation strategy. However, after the second step the GMM estimation is more accurate than the ML estimation. This is probably because the GMM estimation does not need to linearize (5.4). However, the difference is minor, and the time required by the ML estimation is much shorter. The latter holds not only because the GMM needs an additional step, but also because each single step of the GMM estimation takes much more time for the algorithm to converge. Approximately eight hours on a Pentium III computer are required for each step of the GMM estimation, whereas approximately only four hours are needed for the ML estimation.

In figures 5.1 and 5.2, we illustrate the surface of the objective function for our ML estimation. It shows not only the existence of multiple optima but also that the objective function is not smooth. This verifies the necessity of using simulated annealing in our estimation strategy.

Next, we turn to estimating the standard RBC model with U.S. time series data.

5.3 Estimation with Actual Data

5.3.1 The Data Construction

Before estimating the benchmark model with empirical U.S. time series, we first discuss the data that will be used. The empirical studies of RBC models often require a considerable reconstruction of existing macroeconomic data. The time series employed for our estimation should include A_t, the temporary shock in technology; N_t, the labor input (per capita hours); K_t, the capital stock; C_t, consumption; and Y_t, output. Here we can assume that all data except the temporary shock A_t are obtained from statistical sources.

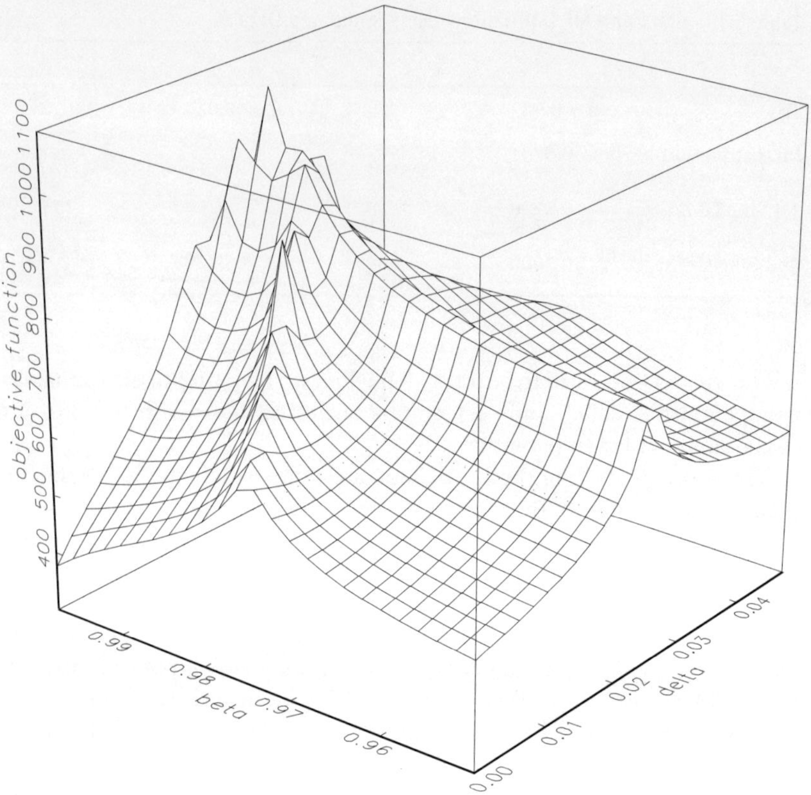

Figure 5.1. The $\beta - \delta$ Surface of the Objective Function for ML Estimation

A common practice is to use the Solow residual for the temporary shock. Assuming that the production function takes the form $Y_t = A_t K_t^{1-\alpha}(N_t X_t)^{\alpha}$, the Solow residual A_t is computed as follows:

$$A_t = \frac{Y_t}{K_t^{1-\alpha}(N_t X_t)^{\alpha}} \tag{5.5}$$

$$= \frac{y_t}{k_t^{1-\alpha} N_t{}^{\alpha}}, \tag{5.6}$$

where X_t follows a constant growth rate:

$$X_t = (1+\gamma)^t X_0. \tag{5.7}$$

Thus, the Solow residual A_t can be derived if the time series Y_t, K_t, N_t; the parameters α and γ; and the initial condition X_0 are given. However,

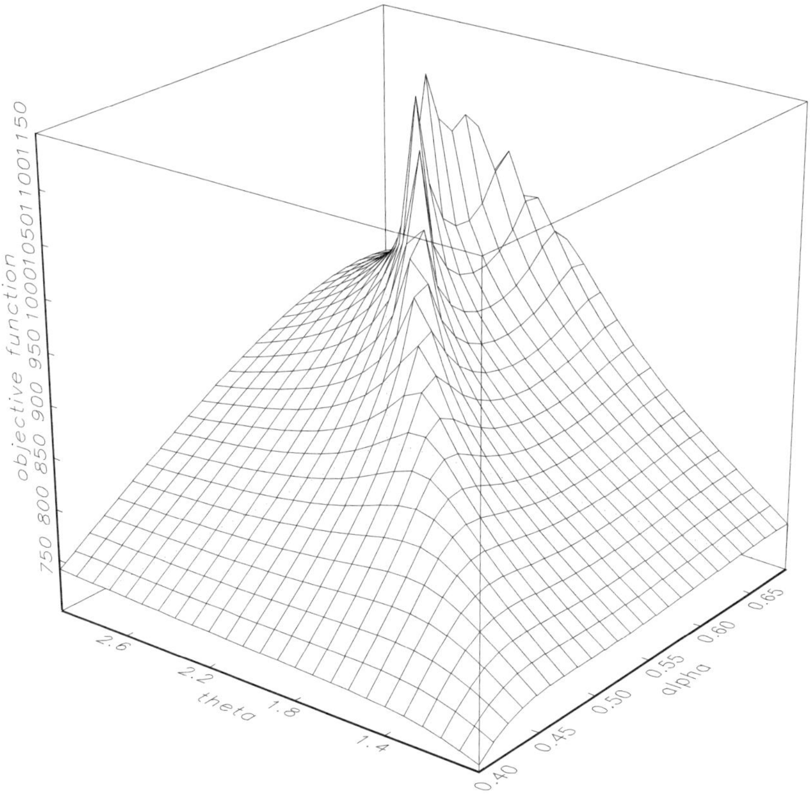

Figure 5.2. The $\theta - \alpha$ Surface of the Objective Function for ML Estimation

deriving the temporary shock as above merits some criticism. This approach uses macroeconomic data that posit full employment, a key assumption in the model that makes it different from other types of models, such as the Keynesian business cycle model. Yet, since this is a common practice, we will follow this procedure. Later in this chapter we will, however, deviate from this practice and construct the Solow residual in a different way. This will allow us to explore what is often called the technology puzzle in the RBC literature.

In addition to constructing the temporary shock, we need to adjust the existing macroeconomic data (such as those from Citibase) to accommodate the definition of the variables in the model.[4] The national income as defined in the model is simply the sum of consumption C_t and investment; the latter increases the capital stock. Therefore, to make the model's national income account consistent with the actual data, one should include government consumption in C_t. Further, it is suggested that not only private investment but also government investment and durable consumption goods (as well as inventory and the value of land) should be included in the capital stock. Consequently, the service generated from durable consumption goods and government capital stock should appear in the definition of Y_t. Since such data are not readily available, one has to compute them based on some assumptions.

Two Different Data Sets

To explore how this treatment of the data construction could affect the empirical assessment of the dynamic optimization model, we will employ two different data sets. Data set I, constructed by Christiano (1987), has been used in many empirical studies of the RBC model, such as Christiano (1988) and Christiano and Eichenbaum (1992). The sample period of this data set is from 1955.1 to 1984.4. All data are quarterly.

Data set II, was obtained mainly from Citibase, except for the capital stock, which was taken from the Current Survey of Business. This data set is taken without any modification. Its sample period is from 1952.1 to 1988.4.

5.3.2 Estimation with the Christiano Data Set

As mentioned, the shock sequence A_t is computed from the time series Y_t, X_t, N_t, and K_t, given the predetermined α, which is denoted as α^* (see equation (5.5)). Here the time series X_t is computed according to equation (5.7) for the given parameter γ and the initial condition X_0. In this estimation, we set X_0 to 1 and γ to 0.0045.[5] Meanwhile, we will consider two α^*'s: one is the standard value 0.58 and the other is 0.66. The latter is the estimated α in Christiano and Eichenbaum (1992). Table 5.2 reports the estimations after the second step of GMM estimation.

All the estimations seem to be quite reasonable. For $\alpha^* = 0.58$, the estimated parameters, though deviating somewhat from the standard parameters used in King et al. (1988a), are all within the economically feasible range. For $\alpha^* = 0.66$, the estimated parameters are very close to those in Christiano and Eichenbaum (1992). Even the parameter β estimated here is very close to the β chosen (rather than estimated) by them. In both cases, the estimated α is almost the same as the predetermined α. This is not surprising because of

Table 5.2. Estimation with Christiano's Data Set

	α	β	δ	θ
$\alpha^* = 0.58$	0.5800 (2.2377E−008)	0.9892 (0.0002)	0.0209 (0.0002)	1.9552 (0.0078)
$\alpha^* = 0.66$	0.6600 (9.2393E−006)	0.9935 (0.0002)	0.0209 (0.0002)	2.1111 (0.0088)

Table 5.3. Estimation with the NIPA Data Set

	α	β	δ	θ
$\alpha^* = 0.58$	0.4656 (71431.409)	0.8553 (54457.907)	0.0716 (89684.204)	1.2963 (454278.06)
$\alpha^* = 0.66$	0.6663 (35789.023)	0.9286 (39958.272)	0.0714 (45174.828)	1.8610 (283689.56)

the way of computing the temporary shocks from the Solow residual. Finally, the standard errors are unusually small.

5.3.3 Estimation with the NIPA Data Set

As in the case of the estimation with Christiano's data set, we again set X_0 to 1 and γ to 0.0045. For the two predetermined α, the estimation for the data set from the National Income and Product Account (NIPA) results in table 5.3.

In contrast to the estimation using the Chrisitano data set, the estimations here are much less satisfying. They deviate significantly from the standard parameters. Some of the parameters, especially β, are not within the economically feasible range. Furthermore, the estimates are all statistically insignificant because of the huge standard errors. Given such a sharp contrast for the results of the two different data sets, one is forced to think about the data issue involved in the current empirical studies of the RBC model. This issue is mostly suppressed in the current debate.

5.4 Calibration and Matching to U.S. Time Series Data

5.4.1 The Data-Generating Process

Given the structural parameters, one can assess the model to see how closely it matches the empirical data. The current method for assessing a stochastic dynamic optimization model of the RBC type is the calibration technique, which was introducod in chapter 3. The basic idea of calibration is to compare the time series moments generated from the model's stochastic simulation with those from a sample economy.

The data generation process for this stochastic simulation is given by the following equations:

$$c_t = G_{11}A_t + G_{12}k_t + g_1 \tag{5.8}$$

$$n_t = G_{21}A_t + G_{22}k_t + g_2 \tag{5.9}$$

$$y_t = A_t k_t^{1-\alpha}(n_t \bar{N}/0.3)^{\alpha} \tag{5.10}$$

$$A_{t+1} = a_0 + a_1 A_t + \epsilon_{t+1} \tag{5.11}$$

$$k_{t+1} = \frac{1}{1+\gamma}\left[(1-\delta)k_t + y_t - c_t\right], \tag{5.12}$$

where $\epsilon_{t+1} \sim N(0, \sigma_\epsilon^2)$; and G_{ij} and g_i $(i, j = 1, 2)$ are the complicated functions of the structural parameters and can be computed from our GAUSS procedure of solving the dynamic optimization problem as presented in appendix II of chapter 1.

The structural parameters used for this stochastic simulation are defined as follows. First, we employ those in table 5.2 at $\alpha^* = 0.58$ for the parameters α, β, δ, and θ. The parameter γ is set 0.0045, as usual. The parameters a_0, a_1 and σ_ϵ in the stochastic equation (5.11) are estimated by the OLS method, given the time series computed from Solow residue.[6] The parameter $\bar{N}$ is the sample mean of per capita hours N_t. For convenience, all these parameters are reported in table 5.4.

Table 5.5 reports our calibration from 5,000 stochastic simulations. In particular, the moment statistics of the sample economy are computed from Christiano's data set, while those for the model economy are generated from our stochastic simulation, using the data generation process (5.8)–(5.12). Here, the moment statistics include the standard deviations of some major macroeconomic variables and their correlation coefficients. For the model economy, we can further obtain the distribution of these moment statistics, which can be reflected by their corresponding standard deviations (in paratheses). The distributions are derived from our 5,000 stochastic simulations. All time series data are detrended by the HP-filter.

5.4.2 The Labor Market Puzzle

In table 5.5, we find that among the four key variables, the volatilities of consumption, capital stock, and output could be regarded as somewhat matched. This is one of the major early results of real business cycle theorists. However, the matching does not hold for employment. The employment in the model economy is excessively smooth.

These results are further demonstrated by figures 5.3 and 5.4, where we compare the observed series from the sample economy against the simulated series with innovation given by the observed Solow residual. The excessive smoothness of employment is a typical problem of the standard model that has been addressed many times in the literature.

Now let us look at the correlations. In the sample economy, there are basically two significant correlations. One is between consumption and output,

Table 5.4. Parameterizing the Standard RBC Model

α	γ	β	δ	θ	$\bar{N}$	a_0	a_1	σ_ϵ
0.5800	0.0045	0.9892	0.0209	1.9552	299.03	0.0333	0.9811	0.0189

Table 5.5. Calibration of Real Business Cycle Model

	Consumption	Capital	Employment	Output
Standard Deviations				
Sample Economy	0.0081	0.0035	0.0165	0.0156
Model Economy	0.0090	0.0037	0.0050	0.0159
	(0.0012)	(0.0007)	(0.0006)	(0.0021)
Correlation Coefficients				
Sample Economy				
Consumption	1.0000			
Capital Stock	0.1741	1.0000		
Employment	0.4604	0.2861	1.0000	
Output	0.7550	0.0954	0.7263	1.0000
Model Economy				
Consumption	1.0000			
	(0.0000)			
Capital Stock	0.2013	1.0000		
	(0.1089)	(0.0000)		
Employment	0.9381	−0.1431	1.0000	
	(0.0210)	(0.0906)	(0.0000)	
Output	0.9796	0.0575	0.9432	1.0000
	(0.0031)	(0.1032)	(0.0083)	(0.0000)

Note: Numbers in parentheses are the corresponding standard deviations.

and the other is between employment and output. Both of these correlations have also been found in our model economy. However, in addition to these two correlations, consumption and employment are significantly correlated in the model economy. In the literature there are different views on the correlation of consumption and employment (see chapter 8). In our context the high correlation should not be surprising, given that in the RBC model the movements of employment and consumption reflect the movements of the same state variables: capital stock and temporary shock. Therefore, they should be correlated. The excessive smoothness of labor effort and the correlation between labor effort and other macrovariables will be taken up in chapter 8.

5.5 The Solow Residual

So far we may argue that one of the major achievements of the standard RBC model is that it could explain the volatility of such key macroeconomic variables as output, consumption, and capital stock. Meanwhile, these results rely on the hypothesis that the driving force of the business cycles is technology shocks, which are assumed to be measured by the Solow residual. The measurement of technology can impact these results in two ways. One is that

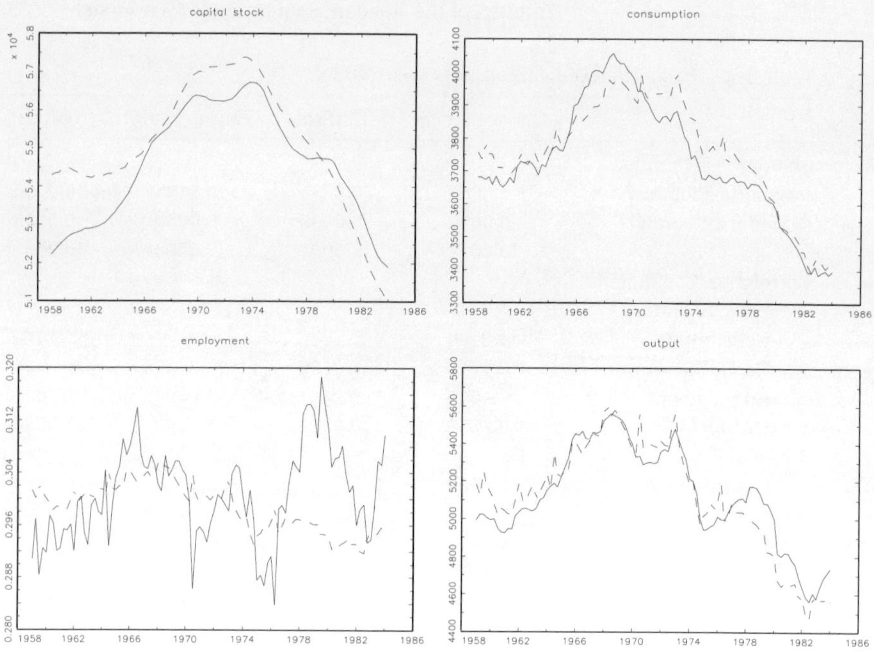

Figure 5.3. Simulated and Observed Series: Nondetrended
Note: Solid line = observed; dashed line = simulated

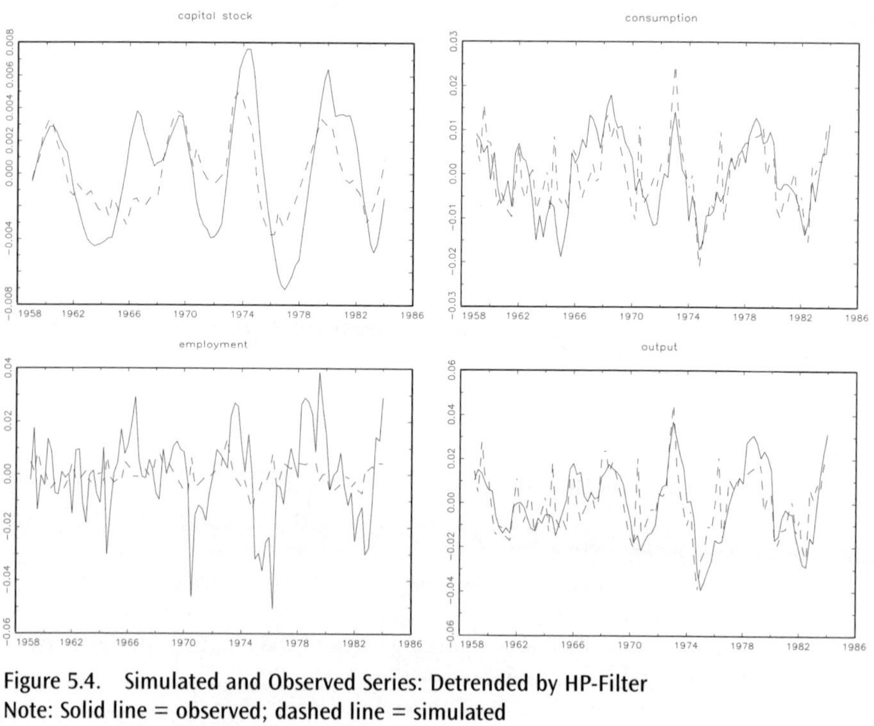

Figure 5.4. Simulated and Observed Series: Detrended by HP-Filter
Note: Solid line = observed; dashed line = simulated

the parameters a_0, a_1, and σ_ϵ in the stochastic equation (5.11) are estimated from the time series computed from the Solow residual. These parameters will directly affect the results of our stochastic simulation. The second is that the Solow residual also serves as the sequence of observed innovations that generate the graphs in figures 5.3 and 5.4. Those innovations are often used in the RBC literature as an additional indicator to support the model and its matching of the empirical data.

Another major presumption of the RBC literature, not shown in table 5.5 but to be shown below, is the technology-driven hypthosis: that technology is procyclical with output, consumption, and employment. This result is also obtained from the empirical evidence, in which the technology is measured by the standard Solow residual.

There are several reasons to distrust the standard Solow residual as a measure of technology shock. First, Mankiw (1989) and Summers (1986) have argued that such a measure often leads to excessive volatility in productivity and even the possibility of technological regress, both of which seem to be empirically implausible. Second, it has been shown that the Solow residual can be expressed by some exogenous variables—for example, demand shocks arising from military spending (Hall 1988) and changed monetary aggregates (Evans 1992)—that are unlikely to be related to factor productivity. Third, the standard Solow residual is not a reliable measure of technology shocks if the cyclical variations in factor utilization are significant.

Considering that the Solow residual cannot be trusted as a measure of the technology shock, researchers have developed different methods to measure technology correctly. All these methods focus on the computation of factor utilization. There are basically three strategies. The first is to use an observed indicator as a proxy for unobserved utilization. A typical example is to employ electricity use as a proxy for capacity utilization (see Burnside et al. 1996). Another strategy is to construct an economic model so that one can compute the factor utilization from the observed variables (see Basu and Kimball 1997; Basu et al. 1998). A third strategy identifies the technology shock through a VAR estimate (see Gali 1999; Francis and Ramey 2001, 2003).

Gali (1999) and Francis and Ramey (2001) have found that if one uses the corrected Solow residual—if one identifies the technology shock correctly—the technology shock is negatively correlated with employment, and therefore the celebrated discovery of the RBC literature must be rejected. Also, if the corrected Solow residual is significantly different from the standard Solow residual, one may find that the standard RBC model, using the Solow residual, can match well the variations in output, consumption, and capital stock, not because the model has been constructed correctly but because it uses a problematic measure of technology.

All these are important problematic issues that are related to the Solow residual. Indeed, if they are confirmed, the RBC model, as driven by technology shocks, may not be a realistic paradigm for macroeconomic analysis. In this section, we will refer to all of this recent research, employing our available data set. We will first follow Hall (1988) and Evans (1992) to test the exogeneity of the Solow residual. In this test, however, we simply use government spending, which is available in Christiano's data set. We will construct a measurement of the technology shock that represents a corrected Solow residual. This construction needs data on factor utilization. Unlike other current research, we use an empirically observed data series, the capacity utilization of manufacturing (IPXMCAQ), obtained from Citibase. Given our new measurement, we then explore whether the RBC model is still able to explain the business cycles, in particular the variation in consumption, output, and capital stock. We will also look at whether the technology still moves procyclically with output, consumption, and employment.

5.5.1 Testing the Exogeneity of the Solow Residual

Apparently, a critical assumption of the Solow residual as a correct measurement of the technology shock is that A_t should be purely exogenous. In other words, the distribution of A_t cannot be altered by the change in other exogenous variables, such as monetary and fiscal policy. Therefore, testing the exogeneity of the Solow residual becomes our first investigation to determine whether the Solow residual is a correct measure of the technology shock.

One possible way to test the exogeneity is to employ the Granger causality test. This is the approach taken by Evans (1992). For this purpose, we will investigate the following specification:

$$A_t = c + \alpha_1 A_{t-1} + \cdots + \alpha_p A_{t-p} + \beta_1 g_{t-1} + \cdots + \beta_p g_{t-p} + \varepsilon_t, \quad (5.13)$$

where g_t is government spending, as an aggregate demand variable, over X_t. If the Solow residual is exogenous, g_t should not have any explanatory power for A_t. Therefore our null hypothesis is

$$H_0 : \beta_1 = \cdots = \beta_p = 0. \quad (5.14)$$

The rejection of the null hypothesis is sufficient for us to refute the assumption that A_t is strictly exogenous. The result of any empirical test for Granger causality can be surprisingly sensitive to the choice of lag length p. The test therefore will be conducted for different lag lengths p's. Table 5.6 provides the corresponding F-statistics computed for the different p's.

The table reveals that at the 5% significance level we can reject the null hypothesis for all the lag lengths p's.[7]

5.5.2 Corrected Technology Shocks

The analysis in section 5.5.1 indicates that the hypothesis that the standard Solow residual be strictly exogenous can be rejected. On the other hand,

Table 5.6. F – Statistics for Testing Exogeneity of Solow Residual

	$F-$ Statistics	Degree of Freedom
$p = 1$	9.5769969	(1, 92)
$p = 2$	4.3041035	(2, 90)
$p = 3$	3.2775435	(4, 86)
$p = 4$	2.3825632	(6, 82)

policy variables—such as government spending, which certainly represents a demand shock—may have explanatory power for the variation of the Solow residual. This finding is consistent with the results in Hall (1988) and Evans (1992). Therefore, we may have sufficient reason to distrust the Solow residual to be a good measure of the technology shock.

Next, we present a simple way of extracting a technology shock from macroeconomic data. If we look at the computation of the Solow residual, such as equation (5.5), we find two strong assumptions inherent in the formulation of the Solow residual. First, it is assumed that the capital stock is fully utilized. Second, it is further assumed that the population follows a constant growth rate that is a part of γ. In other words, there is no variation in population growth. Next we will consider the derivation of the corrected Solow residual by relaxing those strong assumptions.

Let u_t denote the utilization of capital stock, which can be measured by IPXMCAQ from Citibase. The observed output is thus produced by the utilized capital and labor service (expressed in terms of total observed working hours) via the production function:

$$Y_t = \tilde{A}_t \left(u_t K_t\right)^{1-\alpha} \left(Z_t E_t H_t\right)^{\alpha}. \tag{5.15}$$

Above, $\tilde{A}_t$ is the corrected Solow residual (our new measure of temporary shock in technology); E_t is the number of workers employed; H_t denotes the hours per employed worker;[8] and Z_t is the permanent shock in technology. In this formulation, we interpret the utilization of labor service only in terms of working hours, and therefore ignore actual effort, which is more difficult to observe.

Let $\bar{L}_t$ denote the permanent shock to population so that $X_t = Z_t \bar{L}_t$ and L_t denotes the observed population so that $\frac{E_t H_t}{L_t} = N_t$. Dividing both sides of (5.15) by X_t, we obtain

$$y_t = \tilde{A}_t \left(u_t k_t\right)^{1-\alpha} \left(l_t N_t\right)^{\alpha}. \tag{5.16}$$

Above, $l_t \equiv \frac{L_t}{\bar{L}_t}$. Given equation (5.16), the corrected Solow residual $\tilde{A}_t$ can be computed as

$$\tilde{A}_t = \frac{y_t}{\left(u_t k_t\right)^{1-\alpha} \left(l_t N_t\right)^{\alpha}}. \tag{5.17}$$

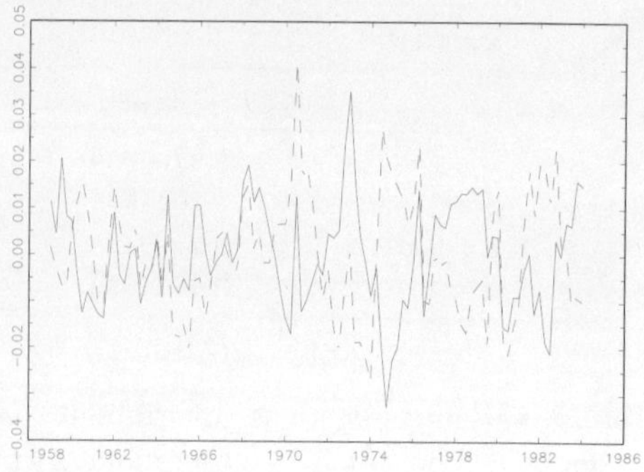

Figure 5.5. The Solow Residual. Standard is solid curve and corrected is dashed curve.

Comparing this with equation (5.5) shows that our corrected Solow residual $\tilde{A}_t$ will match the standard Solow residual A_t if and only if both u_t and l_t equal 1. Figure 5.5 compares these two time series: one for nondetrended, and the other for detrended, series.

As can be observed in figure 5.5, the two series follow basically the same trend and their volatilities are almost the same.[9] However, in the short run, they move in different directions if we compare the detrended series.

5.5.3 Business Cycles with Corrected Solow Residual

Next we use the corrected Solow residual to test the technology-driven hypothesis. In table 5.7, we report the cross-correlations of the technology shock with our four key economic variables: output, consumption, employment, and capital stock. These correlations are compared for three economies: the RBC economy (whose statistics are computed from 5,000 simulations), sample economy I (in which the technology shock is represented by the standard Solow residual), and sample economy II (represented by the corrected Solow residual). The data series are detrended by the HP-filter.

If we look at sample economy I, where the standard Solow residual is employed, we find that the technology shock is procyclical to output, consumption, and employment. This result is exactly predicted by the RBC economy and represents what has been called the technology-driven hypothesis. However, if we use the corrected Solow residual, as in sample economy II,

Table 5.7. The Cross-Correlation of Technology

	Output	Consumption	Employment	Capital Stock
RBC Economy	0.9903	0.9722	0.9966	−0.0255
	(0.0031)	(0.0084)	(0.0013)	(0.1077)
Sample Economy I	0.7844	0.7008	0.1736	−0.2142
Sample Economy II	−.3422	−0.1108	−0.5854	0.0762

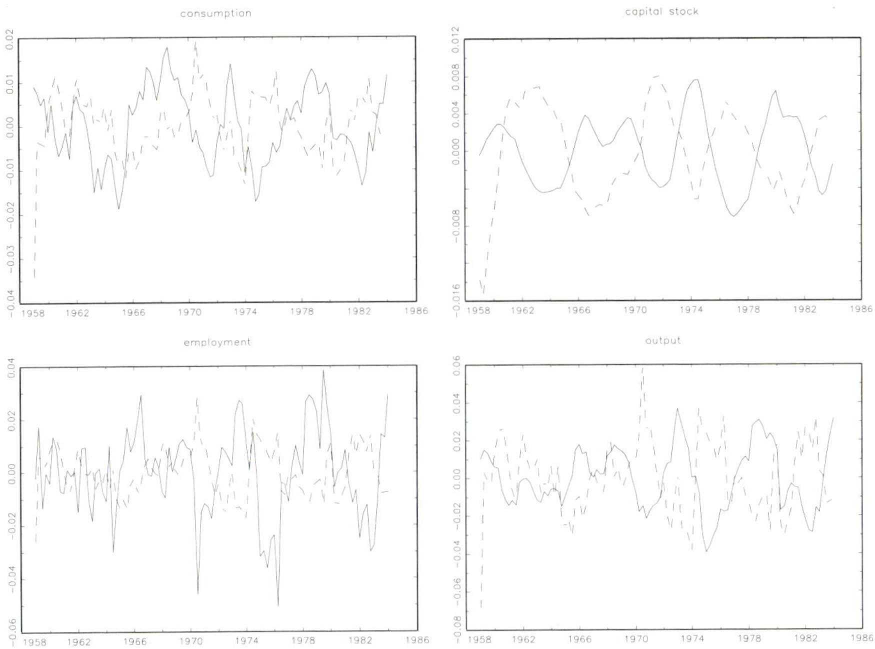

Figure 5.6. Sample and Predicted Moments with Innovation Given by Corrected Solow Residual. Sample is solid line and predicted is dashed line.

we find a somewhat opposite result, especially for employment. We therefore can confirm the findings of the research by Basu et al. (1998), Gali (1999), and Francis and Ramey (2001, 2003).

To test whether the model can still match the observed business cycles, in figure 5.6 we provide a one-time simulation with the observed innovation given by the corrected Solow residual.[10] Comparing figure 5.6 with figure 5.4, we find that the results are in sharp contrast to the prediction in the standard RBC model.

5.6 Conclusions

The standard RBC model has been regarded as a model that replicates the basic moment properties of U.S. macroeconomic time series data despite its rather simple structure. Prescott (1986) summarizes the moment implications thus: “The match between theory and observation is excellent, but far from perfect.” Indeed, many have felt that the RBC research has at least passed the first test. However, this early assessment should be subject to certain qualification.

In the first place, this early assessment builds on the reconstruction of U.S. macroeconomic data. Through its necessity to accommodate the data to the model's implication, such data reconstruction seems to force the first moments of certain macroeconomic variables of the U.S. economy to be matched by the model's steady state at the given economically feasible standard parameters. The unusual small standard errors of the estimates seem to confirm this suspicion.

Second, although one may celebrate the fit of the variation of consumption, output, and capital stock when the reconstructed data series are employed, we still cannot ignore the problems of excessive smoothness of labor effort and the incorrect correlation between labor effort and other macrovariables. These problems are related to the labor market specification of the RBC model. For the model to be able to replicate employment variation, it seems necessary to improve the labor market specification. One possible approach would be to allow for wage stickiness and non-clearing of the labor market, a task that we will turn to in chapter 8.

Third, the fit of the variation in consumption, output, and capital stock may rely on the uncorrected measure of technology. As we have shown in figure 5.6, the match no longer exists when we use the corrected Solow residual as the observed innovations. This incorrect measure of technology takes us to the technology puzzle: The procyclical technology driving the business cycle may not be a very plausible hypothesis. As King and Rebelo (1999, 962) pointed out, "It is the final criticism that the Solow residual is a problematic measure of technology shock that has remained the Achilles heel of the RBC literature." In chapter 9, we will address the technology puzzle again by introducing monopolistic competition into a stochastic dynamic macromodel.

6

Asset Market Implications of Real Business Cycles

6.1 Introduction

In this chapter, we will study asset price implications of the standard RBC model. The idea of employing a basic stochastic growth model to study asset prices goes back to Brock and Mirman (1972) and Brock (1979, 1982). Asset prices contain valuable information about intertemporal decision making, and dynamic models explaining asset pricing are of great importance in current research. Here we want to study a production economy with an asset market and spell out its implications for asset prices and returns. In particular we will explore to what extent it can replicate the empirically found risk-free interest rate, equity premium, and Sharpe ratio.

Modeling asset price and risk premiums in models with production is much more challenging than in exchange economies. Most of the asset pricing literature has followed Lucas (1978) and Mehra and Prescott (1985) in computing asset prices from the consumption-based asset-pricing models with an exogenous dividend stream. Production economies offer a much richer and more realistic environment. First, in economies with an exogenous dividend stream and no savings, consumers are forced to consume their endowment. In economies with production where asset returns and consumption are endogenous, consumers can save, and hence transfer consumption between periods. Second, in economies with an exogenous dividend stream, the aggregate consumption is usually used as a proxy for equity dividends. Empirically, this is not a very sensible modeling choice. Since there is a capital stock in production economies, a more realistic modeling of equity dividends is possible.

Although further extensions of the baseline stochastic growth model of RBC type have been developed to better match actual asset market characteristics,[1] here we will by and large restrict ourselves to the baseline model. The theoretical framework in this chapter is taken from Lettau (1999), Lettau and Uhlig (1999), and Lettau et al. (2001), where the closed-form solutions for risk premiums of equity, long-term real bonds, the Sharpe ratio, and the risk-free interest rates are presented in a log-linearized RBC model developed by Campbell (1994). Those equations can be used as additional moment restrictions in the estimation process. We introduce the asset-pricing restrictions step by step to clearly demonstrate the effect of each new restriction.

First, we estimate the model, using only the restrictions of real variables, as in chapter 5. The data employed for this estimation are taken from Christiano (1987).[2] We then add our first asset-pricing restriction, the risk-free interest rate. We use the observed thirty-day T-bill rate to match the one-period risk-free interest rate implied by the model.[3] The second asset-pricing restriction concerns the risk–return trade-off as measured by the Sharpe ratio, or the price of risk. This variable determines how much expected return agents require per unit of financial risk. Hansen and Jagannathan (1991) and Lettau and Uhlig (1999) show how important the Sharpe ratio[4] is in evaluating asset prices generated by different models. Introducing the Sharpe ratio as a moment restriction in the estimation procedure requires an iterative procedure to estimate the risk aversion parameter. We find that the Sharpe ratio restriction affects the estimation of the model drastically. For each estimation, we compute the implied premiums of equity and long-term real bonds. Those values are then compared with the stylized facts of asset markets.

The estimation technique in this chapter follows the Maximum Likelihood (ML) method, as discussed in chapter 4. All the estimations are made by means of the numerical algorithm, the simulated annealing. In addition, we introduce a diagnostic procedure developed by Watson (1993) and Diebold et al. (1995) to test whether the moments predicted by the model for the estimated parameters can match the moments of the actual macroeconomic time series. In particular, we use the variance-covariance matrix of the estimated parameters to infer the intervals of the moment statistics and to study whether the actual moments derived from the sample data fall within these intervals.

The rest of the chapter is organized as follows. In section 6.2, we use the standard RBC model and log-linearization as proposed by Campbell (1994), and derive the closed-form solutions for the financial variables. Section 6.3 presents the estimation for the model specified by different moments' restrictions. In section 6.4, we interpret our results and contrast the asset market implications of our estimates with the stylized facts of the asset market. Section 6.5 compares the second moments of the time series generated from the model with the moments of actual time series data. Section 6.6 is the conclusion.

6.2 The Standard Model and Its Asset Pricing Implications

6.2.1 The Standard Model

We follow Campbell (1994) and use the notation Y_t for output, K_t for capital stock, A_t for technology, N_t for normalized labor input, and C_t for consumption. The maximization problem of a representative agent is assumed to take the form[5]

$$\max E_t \sum_{i=0}^{\infty} \beta^i \left[\frac{C_{t+i}^{1-\gamma}}{1-\gamma} + \theta \log(1 - N_{t+i}) \right],$$

subject to

$$K_{t+1} = (1-\delta)K_t + Y_t - C_t,$$

with Y_t given by $(A_t N_t)^\alpha K_t^{1-\alpha}$. The first-order condition is given by the following Euler equation:

$$C_t^{-\gamma} = \beta E_t \left\{ C_{t+1}^{-\gamma} R_{t+1} \right\} \tag{6.1}$$

$$\frac{1}{\theta(1-N_t)} = \alpha \frac{A_t^\alpha}{C_t} \left(\frac{K_t}{N_t} \right)^{(1-\alpha)}, \tag{6.2}$$

where R_{t+1} is the gross rate of return on investment in capital, which is equal to the marginal product of capital in production plus undepreciated capital:

$$R_{t+1} \equiv (1-\alpha) \left(\frac{A_{t+1} N_{t+1}}{K_{t+1}} \right)^\alpha + 1 - \delta.$$

We allow firms to issue bonds as well as equity. Since markets are competitive, real allocations will not be affected by this choice (i.e., the Modigliani–Miller theorem is presumed to hold). We denote the leverage factor (the ratio of bonds outstanding to total firm value) as ζ.

At the steady state, technology, consumption, output, and capital stock all grow at a common rate $G = A_{t+1}/A_t$. Hence, (6.1) becomes

$$G = \beta R,$$

where R is the steady state of R_{t+1}. Taking log for both sides, we can further write the above equation as

$$\gamma g = \log(\beta) + r, \tag{6.3}$$

where $g \equiv \log G$ and $r \equiv \log R$. This defines the relation among g, r, β, and γ. In the rest of the chapter, we use g, r, and γ as parameters to be determined; the implied value for the discount factor β can then be deduced from (6.3).

6.2.2 The Log-Linear Approximate Solution

Outside the steady state, the model characterizes a system of nonlinear equations in the logs of technology a_t, consumption c_t, labor n_t, and capital stock k_t. Note that here we use the lowercase letter as the corresponding log variable of the capital letter. In the case of incomplete capital depreciation $\delta < 1$, the exact analytical solution to the model is not feasible. We therefore seek an approximate analytical solution instead. Assume that the technology shock follows an AR(1) process:

$$a_t = \phi a_{t-1} + \varepsilon_t, \tag{6.4}$$

with ε_t the i.i.d. innovation: $\varepsilon_t \sim N(0, \sigma_\varepsilon^2)$. Campbell (1994) shows that the solution, using the log-linear approximation method, can be written as

$$c_t = \eta_{ck} k_t + \eta_{ca} a_t \tag{6.5}$$

$$n_t = \eta_{nk} k_t + \eta_{na} a_t, \tag{6.6}$$

and the law of motion of capital is

$$k_t = \eta_{kk} k_{t-1} + \eta_{ka} a_{t-1}, \tag{6.7}$$

where η_{ck}, η_{ca}, η_{nk}, η_{na}, η_{kk}, and η_{ka} are the complicated functions of the parameters α, δ, r, g, γ, ϕ, and $\bar{N}$ (the steady-state value of N_t).

6.2.3 The Asset Price Implications

The standard RBC model as presented above has strong implications for asset pricing. First, the Euler equation (6.1) implies the following expression regarding the risk-free rate R_t^f:[6]

$$R_t^f = \left\{\beta E_t\left[(C_{t+1}/C_t)^{-\gamma}\right]\right\}^{-1}.$$

Writing the equation in the log form, we obtain the risk-free rate in logs as[7]

$$r_t^f = \gamma E_t \Delta c_{t+1} - \frac{1}{2}\gamma^2 Var \Delta c_{t+1} - \log\beta. \tag{6.8}$$

Using consumption, capital stock, and technology as expressed in (6.5), (6.7), and (6.4) while ignoring the constant term involving the discount factor and the variance of consumption growth, from (6.8) we derive (see Lettau et al. 2001 for the details.):

$$r_t^f = \gamma \frac{\eta_{ck}\eta_{ka}}{1 - \eta_{kk}L}\varepsilon_{t-1}, \tag{6.9}$$

where L is the lag operator. Matching this process implied by the model to the data will give us the first asset market restriction. The second asset market restriction will be the Sharpe ratio, which summarizes the risk–return trade-off:

$$SR_t = \max_{\text{all assets}} \frac{E_t\left[R_{t+1} - R_{t+1}^f\right]}{\sigma_t\left[R_{t+1}\right]}. \tag{6.10}$$

Since the model is log-linear and has normal shocks, the Sharpe ratio can be computed in closed form as:[8]

$$SR = \gamma\eta_{ca}\sigma_\varepsilon. \tag{6.11}$$

Last, we consider the risk premiums of equity (EP) and long-term real bonds (LTBP). These can be computed on the basis of the log-linear solutions (6.5)–(6.7) as

$$LTBP = -\gamma^2\beta\frac{\eta_{ck}\eta_{ka}}{1 - \beta\eta_{kk}}\eta_{ca}^2\sigma_\varepsilon^2 \tag{6.12}$$

$$EP = \left(\frac{\eta_{dk}\eta_{nk} - \eta_{da}\eta_{kk}}{1 - \beta\eta_{kk}} - \gamma\beta\frac{\eta_{ck}\eta_{kk}}{1 - \beta\eta_{kk}}\right)\gamma\eta_{ca}^2\sigma_\varepsilon^2. \tag{6.13}$$

Again we refer the reader to Lettau (1999) and Lettau et al. (2001) for details of those computations.

6.2.4 Some Stylized Facts

Table 6.1 summarizes some key facts of asset markets and real economic activity for the U.S. economy. A successful model should be consistent with these basic moments of real and financial variables. In addition to the well-known stylized facts of macroeconomic variables, we will consider the performance of the model in regard to the following facts of asset markets.

Table 6.1 shows that the equity premium is roughly 2 percent per quarter. The Sharpe ratio, which measures the risk–return trade-off, equals 0.27 in postwar U.S. data. The standard deviations of the real variables reveal the usual hierarchy in volatility, with investment being most volatile and consumption the smoothest. Among the financial variables the equity price

Table 6.1. Asset Market Facts and Real Variables

	Standard Deviation	Mean
GDP	1.72	
Consumption	1.27	
Investment	8.24	
Labor Input	1.59	
T-Bill	0.86	0.19
SP 500	7.53	2.17
Equity Premium	7.42	1.99
Long Bond Premium	0.21	4.80
Sharpe Ratio		0.27

Note: Standard deviations for the real variables are taken from Cooley and Prescott (1995).
The series are H-P filtered.
Asset market data are from Lettau (1999).
All data are from the U.S. economy at quarterly frequency.
Units are percent per quarter.
The Sharpe ratio is the mean of the equity premium divided by its standard deviation.

(SP 500) and equity premium exhibit the highest volatilities, roughly six times higher than consumption.

6.3 The Estimation

6.3.1 The Structural Parameters to Be Estimated

The RBC model presented in section 6.2 contains seven parameters, $\alpha, \delta, r, g, \gamma, \phi$, and $\bar{N}$. The discount factor is determined in (6.3) for given values of g, r, and γ. The parameter θ is dropped because of our log-linear approximation. Of course, we would like to estimate as many parameters as possible, but some of the parameters have to be prespecified. The computation of technology shocks requires the values for α and g. Here we use the standard values of $\alpha = 0.667$ and $g = 0.005$. $\bar{N}$ is specified as 0.3. The parameter ϕ is estimated independently from (6.4) by OLS regression. This leaves the risk aversion parameter γ, the average interest rate r, and the depreciation rate δ to be estimated. The estimation strategy is similar to that of Christiano and Eichenbaum (1992). However, they fix the discount factor and the risk aversion parameter without estimating them. In contrast, the estimation of these parameters is central to our strategy.

6.3.2 The Data

For the real variables of the economy, we use the data set constructed by Christiano (1987). It covers the period from the third quarter of 1955 through the fourth quarter of 1983 (1955.3–1983.4). As we demonstrated in chapter 5, the Christiano data set can match the real side of the economy better than the commonly used NIPA data set. For the time series of the risk-free interest rate, we use the thirty-day T-bill rate to minimize unmodeled inflation risk.

To make the data suitable for estimation, we are required to detrend the data into their log-deviation form. For a data observation X_t, the detrended value x_t is assumed to take the form $\log(X_t/\bar{X}_t)$, where $\bar{X}_t$ is the value of X_t on its steady-state path, that is, $\bar{X}_t = (1+g)^{t-1}\bar{X}_1$. Therefore, for the given g, which can be calculated from the sample, the computation of x_t depends on $\bar{X}_1$, the initial $\bar{X}_t$. We compute this initial condition based on the consideration that the mean of x_t is equal to 0. In other words,

$$\begin{aligned}\frac{1}{T}\sum_{i=1}^{T}\log(X_t/\bar{X}_t) &= \frac{1}{T}\sum_{i=1}^{T}\log(X_t) - \frac{1}{T}\sum_{i=1}^{T}\log(\bar{X}_t)\\ &= \frac{1}{T}\sum_{i=1}^{T}\log(X_t) - \frac{1}{T}\sum_{i=1}^{T}\log(\bar{X}_1) - \frac{1}{T}\sum_{i=1}^{T}\log\left[(1+g)^{t-1}\right]\\ &= 0.\end{aligned}$$

Solving the above equation for $\bar{X}_1$, we obtain

$$\bar{X}_1 = \frac{1}{T}\exp\left\{\sum_{i=1}^{T}\log(X_t) - \sum_{i=1}^{T}\log\left[(1+g)^{t-1}\right]\right\}.$$

6.3.3 The Moment Restrictions of Estimation

For the estimation in this chapter, we use the maximum likelihood (ML) method as discussed in chapter 3. In order to analyze the role of each restriction, we introduce the restrictions step by step. First, we constrain the risk aversion parameter r to unity and use only moment restrictions of the real variables—that is, (6.5)–(6.7)—so we can compare our results with those in Christiano and Eichenbaum (1992). The remaining parameters thus to be estimated are δ and r. We call this model 1 (M1). The matrices for the ML estimation are given by

$$B = \begin{bmatrix} 1 & 0 & 0 \\ -\eta_{ck} & 1 & 0 \\ -\eta_{ck} & 0 & 1 \end{bmatrix}, \quad \Gamma = \begin{bmatrix} -\eta_{kk} & -\eta_{ka} & 0 \\ 0 & 0 & -\eta_{ca} \\ 0 & 0 & -\eta_{na} \end{bmatrix},$$

$$y_t = \begin{bmatrix} k_t \\ c_t \\ n_t \end{bmatrix}, \quad x_t = \begin{bmatrix} k_{t-1} \\ a_{t-1} \\ a_t \end{bmatrix}.$$

After considering the estimation with the moment restrictions only for real variables, we add restrictions from asset markets one by one. We start by including the moment restriction of the risk-free interest rate in estimation while still keeping risk aversion fixed at unity:

$$E\left[b_t - r_t^f\right] = 0,$$

where b_t denotes the return on the thirty-day T-bill and the risk-free rate r_t^f is computed as in (6.9). We refer to this version as model 2 (M2). In this case the matrices B and Γ and the vectors x_t and y_t can be written as

$$B = \begin{bmatrix} 1 & 0 & 0 & 0 \\ -\eta_{nk} & 1 & 0 & 0 \\ -\eta_{nk} & 0 & 1 & 0 \\ 0 & 0 & 0 & 1 \end{bmatrix}, \quad \Gamma = \begin{bmatrix} -\eta_{kk} & -\eta_{ka} & 0 & 0 \\ 0 & 0 & -\eta_{ca} & 0 \\ 0 & 0 & -\eta_{na} & 0 \\ 0 & 0 & 0 & -1 \end{bmatrix},$$

$$y_t = \begin{bmatrix} k_t \\ c_t \\ n_t \\ b_t \end{bmatrix}, \quad x_t = \begin{bmatrix} k_{t-1} \\ a_{t-1} \\ a_t \\ r_t^f \end{bmatrix}.$$

Model 3 (M3) uses the same moment restrictions as model 2 but leaves the risk aversion parameter r to be estimated rather than fixed to unity.

Finally, we impose the condition that the dynamic model should generate a Sharpe ratio of 0.27, as measured in the data (see table 6.1). We take this restriction into account in two different ways. First, as a shortcut, we fix the risk aversion at 50, a value suggested in Lettau and Uhlig (1999) for generating a Sharpe ratio of 0.27, using actual consumption data. Given this value, we estimate the remaining parameters δ and r. This will be called model 4 (M4). In the next version, model 5 (M5), we simultaneously estimate γ and impose a Sharpe ratio restriction of 0.27. Recall from (6.11) that the Sharpe ratio is a function of risk aversion, the standard deviation of the technology shock, and the elasticity of consumption with respect to the shock η_{ca}. Of course, η_{ca} is itself a complicated function of γ. Hence, the Sharpe ratio restriction becomes

$$\gamma = \frac{0.27}{\eta_{\text{ca}}(\gamma)\sigma_\varepsilon}. \tag{6.14}$$

This equation provides the solution of γ, given the parameters δ and r. Since it is nonlinear in γ, we therefore have to use an iterative procedure to obtain the solution. For each given δ and r, searched by the simulated annealing, we first set an initial γ, denoted by γ_0. Then the new γ, denoted

Table 6.2. Summary of Models

Model	Estimated Parameters	Fixed Parameters	Asset Restrictions
M1	r, δ	$\gamma = 1$	none
M2	r, δ	$\gamma = 1$	risk-free rate
M3	r, δ, γ		risk-free rate
M4	r, δ	$\gamma = 50$	risk-free rate, Sharpe ratio
M5	r, δ, γ		risk-free rate, Sharpe ratio

by γ_1, is calculated from (6.14) and is equal to $0.27/[\eta_{ca}(\gamma_0)\sigma_\varepsilon]$. This procedure is continued until convergence.

We summarize the different cases in table 6.2, where we start by using only restrictions on real variables and fix risk aversion to unity (M1). We add the risk-free rate restriction, keeping risk aversion at 1 (M2), then estimate it (M3). Finally we add the Sharpe ratio restriction, fixing risk aversion at 50 (M4), and estimate it by using an iterative procedure (M5). For each model we also compute the implied values of the long-term bond and equity premia, using (6.12) and (6.13).

6.4 The Estimation Results

Table 6.3 summarizes the estimations for the first three models. Standard errors are in parentheses. Entries without standard errors are preset and hence are not estimated.

Consider first model 1, which uses only restrictions on real variables. The depreciation rate is estimated to be just below 2%, close to Christiano and Eichenbaum's (1992) results. The average interest rate is 0.77% per quarter, or 3.08% on an annual basis. The implied discount factor computed from (6.3) is 0.9972. These results confirm the estimates in Christiano and Eichenbaum (1992). Adding the risk-free rate restriction in model 2 does not significantly change the estimates. The discount factor is slightly higher and the average risk-free rate decreases.

However, the implied discount factor now exceeds unity, a problem also encountered in Eichenbaum et al. (1988). Christiano and Eichenbaum (1992)

Table 6.3. Summary of Estimation Results

Model	δ	r	γ
M1	0.0189 (0.0144)	0.0077 (0.0160)	prefixed to 1
M2	0.0220 (0.0132)	0.0041 (0.0144)	prefixed to 1
M3	0.0344 (0.0156)	0.0088 (0.0185)	2.0633 (0.4719)

Note: The standard errors are in parentheses.

avoid this by fixing the discount factor below unity rather than estimating it. Model 3 is more general because the risk aversion parameter is estimated instead of fixed at unity. The ML procedure estimates the risk aversion parameter to be roughly 2—significantly different from 1, the value implied from log-utility function. Adding the risk-free rate restriction increases the estimates of δ and r somewhat. Overall, the model is able to produce sensible parameter estimates when the moment restriction for the risk-free rate is introduced.

While the implications of the dynamic optimization model concerning the real macroeconomic variables could be considered fairly successful, the implications for asset prices are dismal. Table 6.4 computes the Sharpe ratio as well as risk premiums for equity and long-term real bonds, using (6.11)–(6.13). These variables are not used in the estimation of the model parameters. The leverage factor ζ is set to 2/3 for the computation of the equity premium.[9]

Table 6.4 shows that the RBC model is not able to produce sensible asset market prices when the model parameters are estimated from the restrictions derived only from the real side of the model (or, as in M3, adding the risk-free rate). The Sharpe ratio is too small by a factor of 50, and both risk premiums are too small, even negative in certain cases. Introducing the risk-free rate restriction improves the performance only a little. Next, we will try to estimate the model by adding the Sharpe ratio moment restrictions. The estimation is reported in table 6.5.

Model 4 fixes the risk aversion at 50. As explained in Lettau and Uhlig (1999), such a high level of risk aversion has the potential to generate reasonable Sharpe ratios in consumption CAPM models. The question now is how the moment restrictions of the real variables are affected by such a high level of risk aversion. The first row of table 6.5 shows that the resulting estimates are not sensible. The estimates for the depreciation factors and the steady-state interest rate converge to the prespecified constraints,[10] or the estimation does not settle down to an interior optimum. This implies that the real side of the model does not yield reasonable results when risk aversion is 50. High risk aversion implies a low elasticity of intertemporal substitution, so that agents are very reluctant to change their consumption over time.

Trying to estimate risk aversion while matching the Sharpe ratio gives similar results. It is not possible to estimate the RBC model while simultaneously satisfying the moment restrictions from both the real side and the

Table 6.4. Asset Pricing Implications

Model	SR	LT BPrem (%)	EqPrem (%)
M1	0.0065	0.000	−0.082
M2	0.0065	−0.042	−0.085
M3	0.0180	−0.053	−0.091

Table 6.5. Matching the Sharpe Ratio

Model	δ	r	γ
M4	1	0	prefixed to 50
M5	1	1	60

financial side of the model, as shown in the last row in table 6.5. Again the parameter estimates converge to prespecified constraints. The depreciation rate converges again to unity, as does the steady-state interest rate r. The point estimate of the risk aversion parameter is high (60). The reason is, of course, that a high Sharpe ratio requires high risk aversion. The tension between the Sharpe ratio restriction and the real side of the model causes the estimation to fail. It demonstrates again that the asset-pricing characteristics that one finds in the data are fundamentally incompatible with the standard RBC model.

6.5 The Evaluation of Predicted and Sample Moments

Next we provide a diagnostic procedure to compare the second moments predicted by the model with the moments implied by the sample data. Our objective here is to ask whether our RBC model can predict the actual moments of the time series for both the real and the asset market. The moments are revealed by the spectra at various frequencies. A similar diagnostic procedure can be found in Watson (1993) and Diebold et al. (1995).

Given the observations on k_t and a_t and the estimated parameters of our log-linear model, the predicted c_t and n_t can be constructed from the right-hand side of (6.5)–(6.6), with k_t and a_t as their actual observations. We now consider the possible deviations of our predicted series from the sample series, employing our most reasonable estimate, model 3. We can use the variance-covariance matrix of our estimated parameters to infer the intervals of our forecasted series and, hence, the intervals of the moment statistics that we are interested in.

Figure 6.1 presents the Hodrick–Prescott (HP) filtered actual and predicted time series data on consumption, labor effort, risk-free rate, and equity return. As shown in chapter 5, the consumption series can be matched to some degree, whereas the volatility in the labor effort, the risk-free rate, and the equity excess return cannot be matched. The insufficient match of the latter three series is further confirmed by figure 6.2, where we compare the spectra calculated from the data samples with the intervals of the spectra predicted, at 5% significance level, by the models.

A good match of the actual and predicted second moments of the time series would be represented by the solid line falling between the dashed and

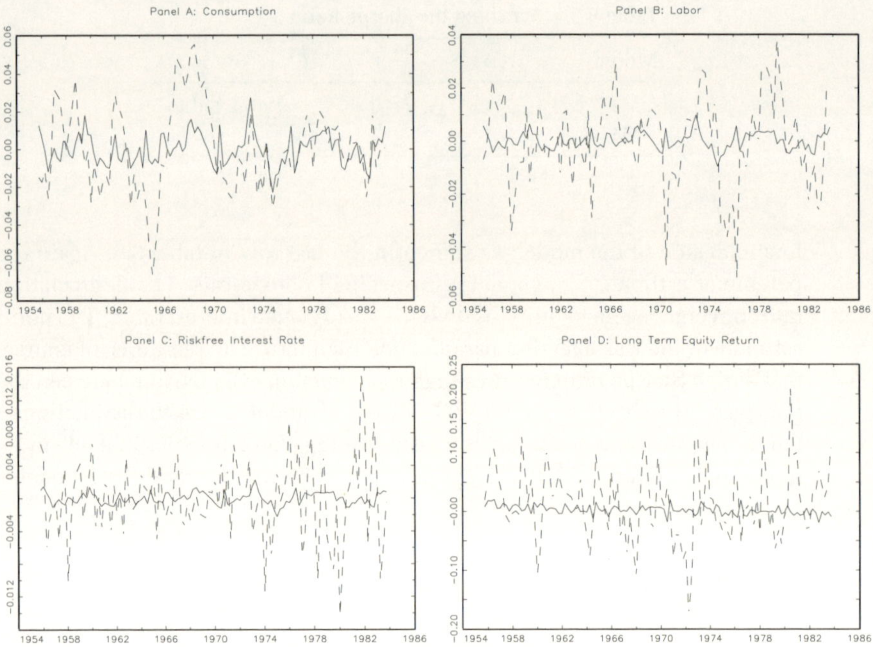

Figure 6.1. Predicted and Actual Series. Note: Solid lines are predicted series, dashed lines are actual series. A = consumption, B = labor, C = risk-free interest rate, D = long-term equity excess return. All variables are HP-detrended except excess equity return.

dotted lines. The time series for labor effort, risk-free interest rate, and equity return fail to do so in figure 6.1.

6.6 Conclusions

Asset prices contain valuable information about intertemporal decision making of economic agents. This chapter has estimated the parameters of a standard RBC model taking the asset-pricing implications into account. We introduced model restrictions based on asset-pricing implications in addition to the standard restrictions of the real variables, and estimated the model by using the ML method. We used the risk-free interest rate and the Sharpe ratio in matching actual and predicted asset market moments, and computed the implicit risk premiums for long-term real bonds and equity. We found that although the inclusion of the risk-free interest rate as a moment restriction can produce sensible estimates, the computed Sharpe ratio and the risk premiums of long-term real bonds and equity were in general counterfactual. The computed Sharpe ratio was too low, and both risk premiums were small, even

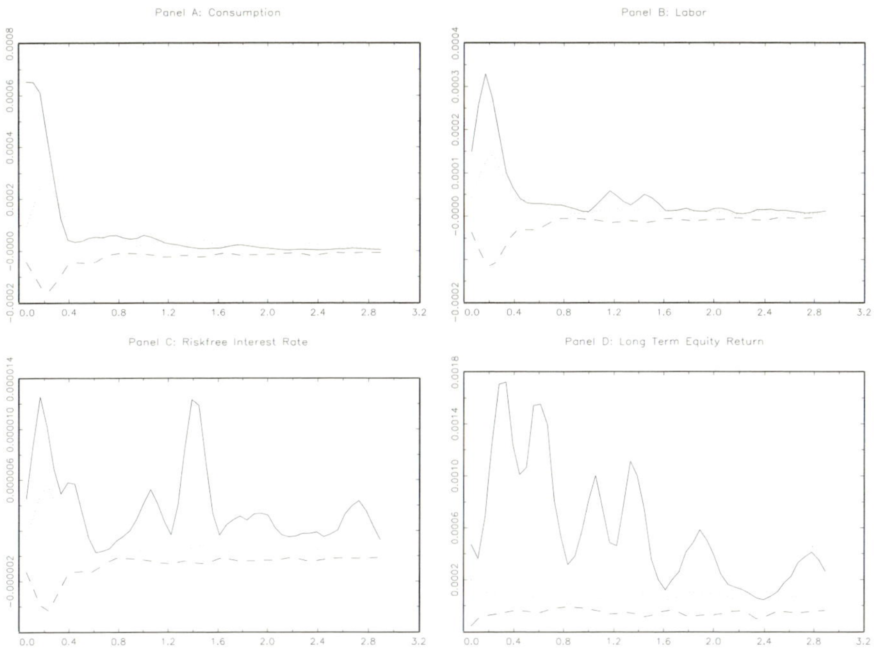

Figure 6.2. The Second-Moment Comparison. Note: Solid line = actual moments; dashed and dotted lines = intervals of predicted moments. A = consumption, B = labor, C = risk-free interest rate, and D = long-term equity excess return. All variables detrended except excess equity return.

negative. Moreover, the attempt to match the Sharpe ratio in the estimation process could hardly generate sensible estimates. Finally, given the sensible parameter estimates, the second moments of labor effort, risk-free interest rate, and long-term equity return predicted by the model did not match the corresponding moments of the sample economy well.

We concluded that the standard RBC model cannot match the asset market restrictions, at least with the standard technology shock, constant relative risk aversion utility function, and no adjustment costs. Other researchers have looked at extensions of the standard model, such as technology shocks with a greater variance, other utility functions (e.g., utility functions with habit formation), and adjustment costs of investment. The latter line of research has been pursued by Jerman (1998) and Boldrin et al. (1996, 2001).[11] Those extensions of the standard model are, at least to a certain extent, more successful in replicating stylized asset market characteristics, yet they frequently use extreme parameter values in order to match the asset price characteristics of the model with the data. Moreover, the approximation methods for solving the models may not be very reliable because accuracy tests for the approximation methods used are still missing.[12]

PART III

BEYOND THE STANDARD MODEL

Model Variants with Keynesian Features

7

Multiple Equilibria and History Dependence

7.1 Introduction

One of the important features of Keynesian economics is that there is no unique equilibrium toward which the economy moves. The dynamics are open ended in the sense that it can move to a low level or a high level of economic activity and expectations, and policy may become important in tilting the dynamics to one outcome or the other.[1]

In recent times such dynamics have been found in a large number of dynamic models with intertemporal optimization. Those models have been called indeterminacy models. Theoretical models of this type are reviewed in Benhabib and Farmer (1999) and Farmer (1999), and an empirical assessment is given in Schmidt-Grohe (2001). Some of the models are real models, RBC models, with increasing return to scale and/or more general preferences (as introduced in chapter 4 and 5) that can exhibit locally stable steady-state equilibria giving rise to sunspot phenomena.[2] Multiple equilibria can also arise as a consequence of increasing returns to scale and/or more general preferences. Other models are monetary ones, where consumers' welfare is affected positively by consumption and cash balances, and negatively by the labor effort and an inflation gap from certain target rates. For certain substitution properties between consumption and cash holdings, those models admit both unstable and stable high-level and low-level steady states. Here there can be indeterminacy in the sense that any initial condition in the neighborhood of one of the steady states is associated with a path toward or away from that steady state (see Benhabib et al. 2001).

When indeterminacy models exhibit multiple steady-state equilibria, and one of the equilibria is an attractor (repeller), this permits any path in the vicinity of this steady-state equilibrium to move toward (away from) that steady-state equilibrium. Despite some unresolved issues in the literature on multiple equilibria and indeterminacy,[3] this type of work has greatly enriched macrodynamic modeling.

Pursuing this line of research, we show that one does not need to refer to increasing returns to scale or specific preferences to obtain such results. Through the adjustment cost of capital we may obtain non-uniqueness of steady-state equilibria in an otherwise standard dynamic optimization version. Multiple steady-state equilibria, in turn, may lead to thresholds separating different domains of attraction of capital stock, consumption, employment, and welfare level. Thresholds are important as separation points below or above which it is advantageous to move to lower or higher levels of capital stock, consumption, employment, and welfare. Our model thus can explain how the economy becomes history dependent and moves, after a shock or policy influences, to low- or high-level equilibria in employment and output.

Recently, numerous stochastic growth models have employed adjustment cost of capital. In nonstochastic dynamic models, adjustment cost has been used by Eisner and Stroz (1963), Lucas (1967), and Hayashi (1982). Authors in this tradition have also distinguished the absolute adjustment cost depending on the level of investment from the adjustment cost depending on investment relative to capital stock (Uzawa 1968; Asada and Semmler 1995).[4] In stochastic growth models, adjustment cost has been used by Boldrin et al. (2001), and adjustment cost associated with the rate of change of investment can be found in Christiano et al. (2001). In this chapter we show that adjustment cost in a standard RBC model can give rise to multiple steady-state equilibria. The existence of multiple steady-state equilibria entails thresholds that separate different domains of attraction for welfare and employment and allow for an open-ended dynamics depending on the conditions and policy influences impacting the initial conditions.

The remainder of this chapter is organized as follows. Section 7.2 presents the model. Section 7.3 studies the adjustment cost function that gives rise to multiple equilibria, and section 7.4 demonstrates the existence of thresholds[5] that separate domains of attraction. Section 7.5 concludes the chapter. The proof of the propositions in the text is provided in the appendix.

7.2 The Model

The model we present here is the standard stochastic growth model of the RBC type, as in King et al. (1988a), augmented by adjustment cost. The state equation for the capital stock takes the form

$$K_{t+1} = (1 - \delta)K_t + I_t - Q_t, \tag{7.1}$$

where

$$I_t = Y_t - C_t \tag{7.2}$$

and

$$Y_t = A_t K_t^{1-\alpha}(N_t X_t)^{\alpha}. \tag{7.3}$$

Above, K_t, Y_t, I_t, C_t, and Q_t are, respectively, the level of capital stock, output, investment, consumption, and adjustment cost, all in real terms. N_t is per capita working hours; A_t is the temporary shock in technology; and X_t is the permanent (including both population and productivity growth) shock that follows a growth rate γ. The model is nonstationary due to X_t. To transform the model into a stationary version, we need to detrend the variables. For this purpose, we divide both sides of equations (7.1)–(7.3) by X_t:

$$k_{t+1} = \frac{1}{1+\gamma}\left[(1-\delta)k_t + i_t - q_t\right]$$

$$i_t = y_t - c_t \tag{7.4}$$

$$y_t = A_t k_t^{1-\alpha}(n_t \bar{N}/0.3)^{\alpha}. \tag{7.5}$$

Above, k_t, c_t, i_t, y_t, and q_t are the detrended variables for K_t, C_t, I_t, Y_t, and Q_t[6]; n_t is defined to be $\frac{0.3N_t}{\bar{N}}$, with $\bar{N}$ denoting the sample mean of N_t. Here n_t is often regarded to be the normalized hours with its sample mean equal to 30%. We will assume that the detrended adjustment cost q_t depends on detrended investment i_t:

$$q_t = q(i_t).$$

The objective function takes the form

$$\max E_0 \sum_{t=0}^{\infty} \beta^t \left[\log c_t + \theta \log(1 - n_t)\right].$$

To solve the model, we first form the Lagrangian:

$$L = \sum_{t=0}^{\infty} \beta^t \left[\log(c_t) + \theta \log(1-n_t)\right]$$
$$- \sum_{t=0}^{\infty} E_t \left\{\beta^{t+1}\lambda_{t+1}\left[k_{t+1} - \frac{1}{1+\gamma}\left[(1-\delta)k_t + i_t - q(i_t)\right]\right]\right\}.$$

Setting to 0 the derivatives of L with respect to c_t, n_t, k_t, and λ_t, we obtain the following first-order conditions:

$$\frac{1}{c_t} - \frac{\beta}{1+\gamma} E_t \lambda_{t+1} \left[1 - q'(i_t)\right] = 0 \tag{7.6}$$

$$\frac{-\theta}{1-n_t} + \frac{\beta}{1+\gamma} E_t \lambda_{t+1} \frac{\alpha y_t}{n_t} \left[1 - q'(i_t)\right] = 0 \tag{7.7}$$

$$\frac{\beta}{1+\gamma} E_t \lambda_{t+1} \left[(1-\delta) + \frac{(1-\alpha) y_t}{k_t} \left[1 - q'(i_t)\right] \right] = \lambda_t \tag{7.8}$$

$$k_{t+1} = \frac{1}{1+\gamma} \left[(1-\delta) k_t + i_t - q(i_t) \right], \tag{7.9}$$

with i_t and y_t to be given by (7.4) and (7.5), respectively.

The following proposition concerns the steady states.

Proposition 5 *Assume A_t has a steady state $\bar{A}$. Equations (7.4)–(7.9), when evaluated at their certainty equivalence form, determine the following steady states:*

$$[b\phi(\bar{i}) - 1][\bar{i} - q(\bar{i})] - a\phi(\bar{i})^{1-\frac{1}{\alpha}} - q(\bar{i}) = 0 \tag{7.10}$$

$$\bar{k} = \frac{1}{\gamma + \delta} [\bar{i} - q(\bar{i})] \tag{7.11}$$

$$\bar{n} = \left[\frac{\phi(\bar{i})}{\bar{A}} \right]^{\frac{1}{\alpha}} \bar{k} \frac{0.3}{\bar{N}} \tag{7.12}$$

$$\bar{y} = \phi(\bar{i}) \bar{k} \tag{7.13}$$

$$\bar{c} = \bar{y} - \bar{i} \tag{7.14}$$

$$\bar{\lambda} = \frac{(1+\gamma)}{\beta \bar{c} \left[1 - q'(\bar{i})\right]}. \tag{7.15}$$

where

$$a = \frac{\alpha}{\theta} \bar{A}^{\frac{1}{\alpha}} (\bar{N}/0.3) \tag{7.16}$$

$$b = \frac{(1 + \frac{\alpha}{\theta})}{\gamma + \delta} \tag{7.17}$$

$$\phi(\bar{i}) = \frac{m}{1 - q'(\bar{i})} \tag{7.18}$$

and

$$m = \frac{(1+\gamma) - (1-\delta)\beta}{\beta(1-\alpha)}. \tag{7.19}$$

Equation (7.10) determines the solution of $\bar{i}$, depending on the assumption that all other steady states can be uniquely solved via (7.11)–(7.15). Also, if $q(i)$ is linear, (7.18) indicates that $\phi(\cdot)$ is constant, and then $\bar{i}$ is uniquely determined from (7.10). Therefore, if $q(i)$ is linear, no multiple steady-state equilibria will occur.

7.3 The Existence of Multiple Steady States

Many nonlinear forms of $q(i)$ may lead to a multiplicity of equilibria. Here we only consider that $q(i)$ takes the logistic form

$$q(i) = \frac{q_0 \exp(q_1 i)}{\exp(q_1 i) + q_2} - \frac{q_0}{1 + q_2}. \tag{7.20}$$

Figure 7.1 shows a typical shape of $q(i)$, while figure 7.2 shows the corresponding derivatives $q'(i)$ with the parameters given in table 7.1.

In equation (7.20), we posit a restriction such that $q(0) = 0$. Another restriction, which is reflected in figure 7.1, is that

$$q(i) < i, \tag{7.21}$$

indicating that the adjustment cost should never be larger than the investment itself. Both restrictions seem reasonable.

The critical points, i_1 and i_2, in figures 7.1 and 7.2 need to be discussed. These are the points at which $q'(i) = 1$. Between i_1 and i_2, $q'(i) > 1$. When $q'(i) > 1$, equation (7.18) indicates that $\phi(i)$ is negative because from (7.19) $m > 0$. A negative $\phi(i)$ will lead to a complex $\phi(i)^{1-\frac{1}{\alpha}}$ in (7.10). We therefore

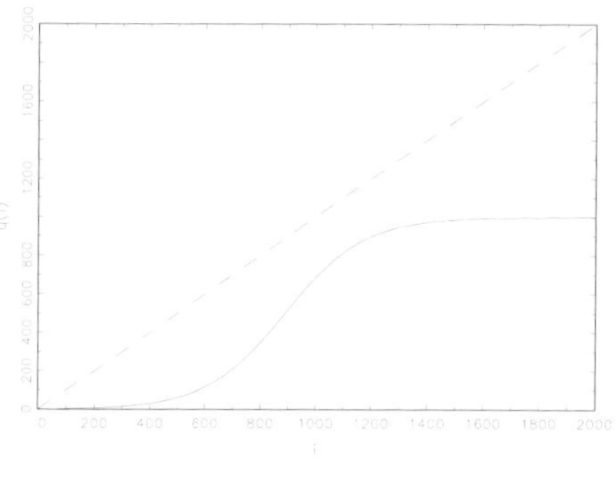

Figure 7.1.
The Adjustment Cost Function

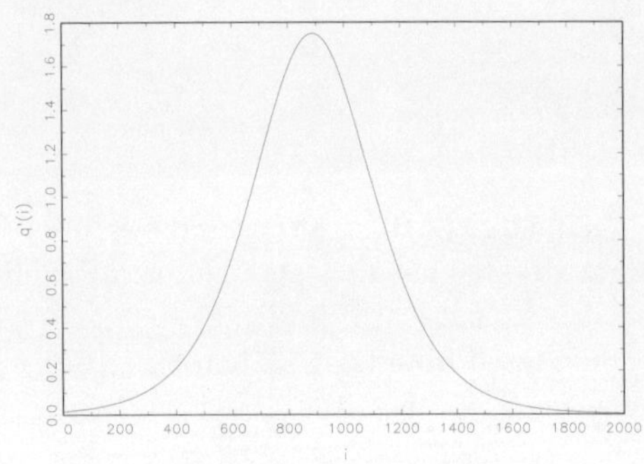

Figure 7.2. The Derivatives of the Adjustment Cost

obtain two feasible ranges for the existence of steady states of i : one is $(0, i_1)$ and the other is $(i_2, +\infty)$.

The following proposition concerns the existence of multiple equilibria.

Proposition 6 *Let $f(i) \equiv [b\phi(i) - 1][i - q(i)] - a\phi(i)^{1-\frac{1}{\alpha}} - q(i)$, where $q(i)$ takes the form as in (7.20), subject to (7.21). Assume that $bm - 1 > 0$.*

- *There exists one and only one i, denoted $\bar{i}_1$, in the range $(0, i_1)$ such that $f(i) = 0$.*
- *In the range $(i_2, +\infty)$, if there are some i's at which $f(i) < 0$, then there must exist two i's, denoted as $\bar{i}_2$ and $\bar{i}_3$, such that $f(i) = 0$.*

The assumption $bm - 1 > 0$ is plausible, given the standard parameters for b and m; also, $f(i) = 0$ is indeed equation (2.10). Therefore, this proposition indicates a condition under which multiple steady states will occur. In particular, if there exist some i's in $(i_2, +\infty)$ at which $f(i) < 0$, three equilibria will occur. A formal mathematical proof of the existence of this condition is intractable. In figure 7.3, we show the curve of $f(\cdot)$, given the empirically plausible parameters reported in table 7.1 and other standard parameters given in table 7.2. The curves cut the 0 line three times, indicating three steady states of i.

We use a numerical method to compute the three steady states of i: $\bar{i}_1$, $\bar{i}_2$, and $\bar{i}_3$. Given these steady states, the other steady states are computed by (7.11)–(7.15). Table 7.3 uses essentially the same parameters as reported in table 5.1, and shows the results of the computations of the three steady states.

Table 7.1. The Parameters in the Logistic Function

q_0	q_1	q_2
2,500	0.0034	500

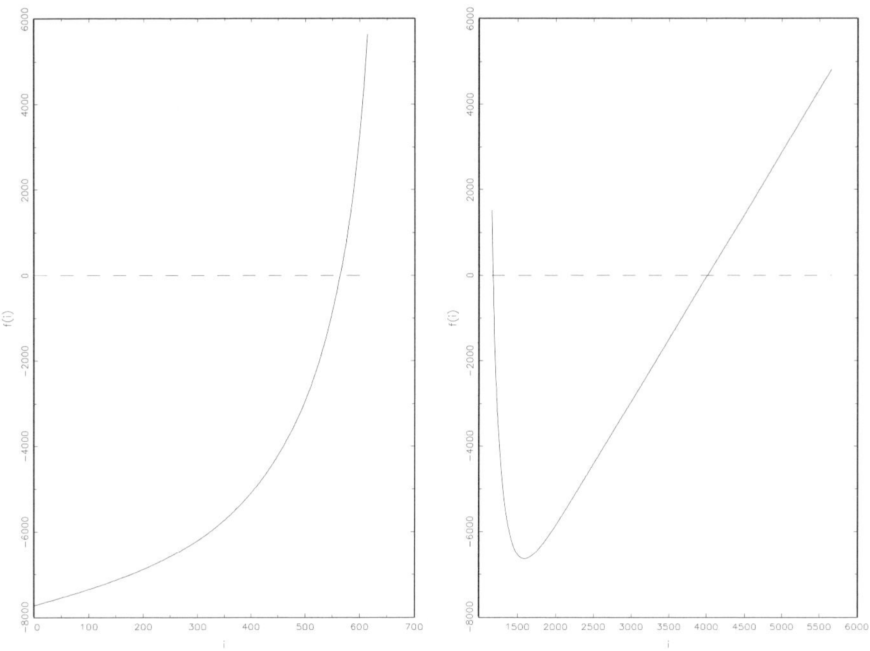

Figure 7.3. Multiplicity of Equilibria: $f(i)$ Function

Above, $\bar{V}$ is the value of the objective function at the corresponding steady states. It therefore reflects one corresponding welfare level. The steady state corresponding to $\bar{i}_2$ deserves some discussion. The welfare of $\bar{i}_1$ is larger than that of $\bar{i}_2$; the corresponding steady states of $\bar{i}_1$ in capital, output, and consumption are all greater than those corresponding to $\bar{i}_2$; yet the corresponding steady state in labor effort, and thus employment, is larger for $\bar{i}_2$. This indicates that $\bar{i}_2$ may be inferior in terms of welfare, at least compared with $\bar{i}_1$. On the other hand $\bar{i}_1$ and $\bar{i}_3$ also exhibit differences in welfare and employment.

Table 7.2. The Standard Parameters of RBC the Model

α	γ	β	δ	θ	$\bar{N}$	$\bar{A}$
0.5800	0.0045	0.9930	0.2080	2.0189	480.00	1.7619

Note: $\bar{N}$ is calculated on the assumption of 12 weeks per quarter and 40 working hours per week. $\bar{A}$ is derived from $A_t = a_0 + a_1 A_{t-1} + \varepsilon_t$, with estimated a_0 and a_1 given by 0.0333 and 0.9811, respectively.

Table 7.3. The Multiple Steady States

	Corresponding to $\bar{i}_1$	Corresponding to $\bar{i}_2$	Corresponding to $\bar{i}_3$
$\bar{i}$	564.53140	1175.7553	4010.4778
$\bar{k}$	18667.347	11672.642	119070.11
$\bar{n}$	0.2543659	0.3090471	0.3378172
$\bar{c}$	3011.4307	2111.2273	5169.5634
$\bar{y}$	3575.9621	3286.9826	9180.0412
$\bar{\lambda}$	0.0008346	0.0017500	0.0001956
$\bar{V}$	1058.7118	986.07481	1101.6369

7.4 The Solution

An analytical solution to the dynamics of the model with adjustment cost is not feasable. We therefore have to rely on an approximate solution. For this, we first linearize the first-order conditions around the three sets of steady states as reported in table 7.3. Then, by applying an approximation method as discussed in chapter 2, we obtain three sets of linear decision rules for c_t and n_t corresponding to our three sets of steady states. For notational convenience, we denote them as decision rule set 1, set 2 and set 3, corresponding to $\bar{i}_1$, $\bar{i}_2$, and $\bar{i}_3$. Assume that A_t stays at its steady state $\bar{A}$ so that we consider only the deterministic case. The ith set of decision rules can then be written as

$$c_t = G_c^i k_t + g_c^i \tag{7.22}$$

$$n_t = G_n^i k_t + g_n^i, \tag{7.23}$$

where $i = 1, 2, 3$. We therefore can simulate the solution paths by using equations (7.22) and (7.23) together with (7.4), (7.5), and (7.9).

The question then arises as to which set of decision rules, as expressed by (7.22) and (7.23), should be used. The likely conjecture is that this will depend on the initial condition k_0. For example, if k_0 is close to $\bar{k}_1$, the steady state of k corresponding to $\bar{i}_1$, we would expect that decision rule set 1 is appropriate. This consideration further indicates that there must exist some thresholds for k_0 at which intervals are marked regarding which set of decision rules should be applied. To detect such thresholds, we compute the values of the objective functions starting at different k_0 for our three decision rules. Specifically, we compute V, where

$$V = \sum_{t=0}^{\infty} \beta^t [\log (c_t) + \theta \log (1 - n_t)].$$

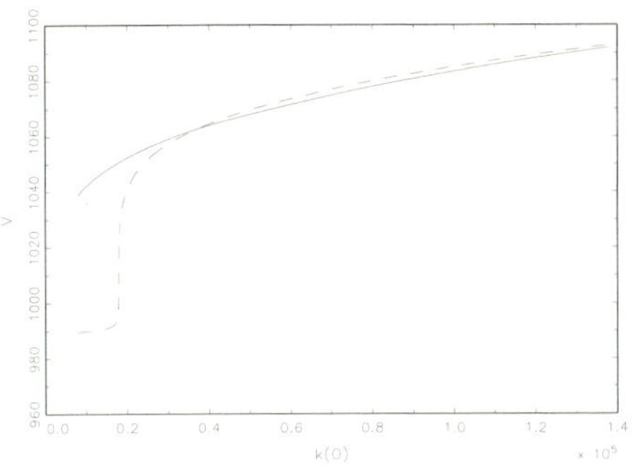

Figure 7.4. The Welfare Performance of Three Linear Decision Rules. Note: Solid line is rule set 1; dotted line is rule set 2; dashed line is rule set 3.

We should choose the range of k_0's that covers the three steady states of $\bar{k}$'s as reported in table 7.3. In this exercise, we choose the range [8,000, 138,000] for k_0. Figure 7.4 compares the welfare performance of our three sets of linear decision rules.

From figure 7.4, we realize that the value of the objective function is always lower for decision rule set 2. This is likely to be caused by its inferior welfare performance at the steady states for which we compute the decision rule. However, there is an intersection of the two welfare curves corresponding to decision rule sets 1 and 3. This intersection, occurring around $k_0 =$ 36,900, can be regarded as a threshold. If $k_0 <$ 36,900, the household should choose decision rule set 1 because it will allow the household to obtain a higher welfare. On the other hand, if $k_0 >$ 36,900, the household may choose decision rule set 3, since this leads to a higher welfare.

7.5 Conclusion

This chapter shows that the introduction of adjustment cost of capital may lead to non-uniqueness of steady-state equilibria in an otherwise standard RBC model. Multiple steady-state equilibria, in turn, lead to thresholds separating the domains of attraction of capital stock, consumption, employment, and welfare level. As our simulation shows, thresholds are important as separation points below or above which it is optimal to move to lower or higher levels of capital stock, consumption, employment, and welfare. Our model thus can easily explain how an economy becomes history dependent and moves, after a shock, to a low- or high-level equilibrium in employment and output. Further economic models giving rise to multiple equilibria and thresholds are presented in Grüne and Semmler (2004a).

The above model stays as close as possible to the standard RBC model except that, for illustrative purposes, a specific form of adjustment cost of capital is introduced. On the other hand, dynamic models giving rise to indeterminacy usually have to presume some weak externalities and increasing returns and/or more general preferences. Kim (2004) discusses to what extent weak externalities in combination with more complex preferences will produce indeterminacy. He shows that if for the generalized RBC model (see chapter 4, section 4.5) there is a weak externality, $\xi > 0$, then the model generates local indeterminacy. Overall, as shown in many recent contributions, the issue of multiple equilibria and indeterminacy is an important macroeconomic issue and should be pursued in future research.

7.6 Appendix: The Proof of Propositions 5 and 6

7.6.1 The Proof of Proposition 5

The certainty equivalence of equations (7.4)–(7.9) takes the form

$$\frac{1}{\bar{c}} - \frac{\beta}{1+\gamma}\bar{\lambda}\left[1 - q'(\bar{i})\right] = 0 \tag{7.24}$$

$$\frac{-\theta}{1-\bar{n}} + \frac{\beta}{1+\gamma}\bar{\lambda}\frac{\alpha\bar{y}}{\bar{n}}\left[1 - q'(\bar{i})\right] = 0 \tag{7.25}$$

$$\frac{\beta}{1+\gamma}\left[(1-\delta) + \frac{(1-\alpha)\bar{y}}{\bar{k}}\left[1 - q'(\bar{i})\right]\right] = 1 \tag{7.26}$$

$$\bar{k} = \frac{1}{1+\gamma}\left[(1-\delta)\bar{k} + \bar{i} - q(\bar{i})\right] \tag{7.27}$$

$$\bar{i} = \bar{y} - \bar{c} \tag{7.28}$$

$$\bar{y} = \bar{A}\bar{k}^{1-\alpha}(\bar{n}\bar{N}/0.3)^{\alpha}. \tag{7.29}$$

From (7.27),

$$\bar{k} = \frac{1}{\gamma+\delta}\left[\bar{i} - q(\bar{i})\right], \tag{7.30}$$

which is equation (7.11). From (7.26),

$$\begin{aligned}\frac{\bar{y}}{\bar{k}} &= \frac{(1+\gamma) - (1-\delta)\beta}{\beta(1-\alpha)\left[1 - q'(\bar{i})\right]} \\ &= \phi(\bar{i}),\end{aligned} \tag{7.31}$$

which is equation (7.13) with $\phi(\bar{i})$ given by (7.18) and (7.19). Further, from (7.29),

$$\frac{\bar{y}}{\bar{k}} = \bar{A}\bar{k}^{-\alpha}(\bar{n}\,\bar{N}/0.3)^{\alpha}. \tag{7.32}$$

Using (7.31) to express $\frac{\bar{y}}{\bar{k}}$, from (7.32) we derive

$$\bar{n} = \left[\frac{\phi(\bar{i})}{\bar{A}}\right]^{\frac{1}{\alpha}} \bar{k}\frac{0.3}{\bar{N}}, \tag{7.33}$$

which is (7.12). Next, using (7.28) to express $\bar{i}$ while using $\phi(\bar{i})\bar{k}$ for $\bar{y}$, from (7.30) we derive $(\gamma+\delta)\bar{k} = \phi(\bar{i})\bar{k} - \bar{c} - q(\bar{i})$, which is equivalent to

$$\begin{aligned}\bar{c} &= (\phi(\bar{i}) - \gamma - \delta)\bar{k} - q(\bar{i})\\ &= c_1(\phi(\bar{i}), \bar{k}, q(\bar{i})).\end{aligned} \tag{7.34}$$

Meanwhile, from (7.24) and (7.25)

$$\frac{\beta}{1+\gamma}\bar{\lambda}\left[1 - q'(\bar{i})\right] = \frac{1}{\bar{c}} = \frac{\theta}{(1-\bar{n})\alpha}\left(\frac{\bar{y}}{\bar{n}}\right)^{-1}.$$

The first equation is equivalent to (7.15), and the second equation indicates

$$\bar{c} = \frac{(1-\bar{n})\alpha}{\theta}\left(\frac{\bar{y}}{\bar{n}}\right), \tag{7.35}$$

where, from (7.29),

$$\begin{aligned}\frac{\bar{y}}{\bar{n}} &= \bar{A}\left(\frac{\bar{k}}{\bar{n}}\right)^{1-\alpha}(\bar{N}/0.3)^{\alpha}\\ &= \bar{A}\left(\frac{\bar{y}}{\bar{n}}\right)^{1-\alpha}\left(\frac{\bar{y}}{\bar{k}}\right)^{\alpha-1}(\bar{N}/0.3)^{\alpha}\\ &= \bar{A}^{\frac{1}{\alpha}}\phi(\bar{i})^{1-\frac{1}{\alpha}}(\bar{N}/0.3).\end{aligned} \tag{7.36}$$

Substitute (7.36) into (7.35) and then express $\bar{n}$ in terms of (7.33):

$$\begin{aligned}\bar{c} &= \frac{(1-\bar{n})\alpha}{\theta}\bar{A}^{\frac{1}{\alpha}}\phi(\bar{i})^{1-\frac{1}{\alpha}}(\bar{N}/0.3)\\ &= \frac{\alpha}{\theta}\bar{A}^{\frac{1}{\alpha}}(\bar{N}/0.3)\phi(\bar{i})^{1-\frac{1}{\alpha}} - \frac{\alpha}{\theta}\bar{A}^{\frac{1}{\alpha}}(\bar{N}/0.3)\phi(\bar{i})^{1-\frac{1}{\alpha}}\left[\frac{\phi(\bar{i})}{\bar{A}}\right]^{\frac{1}{\alpha}}\bar{k}\frac{0.3}{\bar{\mathrm{N}}}\\ &= a\phi(\bar{i})^{1-\frac{1}{\alpha}} - \frac{\alpha}{\theta}\phi(\bar{i})\bar{k}\\ &= c_2(\phi(\bar{i}), \bar{k}),\end{aligned}$$

$$= c_2(\phi(\bar{i}), \bar{k}), \tag{7.37}$$

with a given by (7.16).

Let $c_1(\cdot) = c_2(\cdot)$. We thus obtain

$$(\phi(\bar{i}) - \gamma - \delta)\bar{k} - q(\bar{i}) = a\phi(\bar{i})^{1-\frac{1}{\alpha}} - \frac{\alpha}{\theta}\phi(\bar{i})\bar{k},$$

which is equivalent to

$$\left[(1 + \frac{\alpha}{\theta})\phi(\bar{i}) - \gamma - \delta\right]\bar{k} = a\phi(\bar{i})^{1-\frac{1}{\alpha}} + q(\bar{i}).$$

Using (7.30) for $\bar{k}$, we obtain

$$\left[(1 + \frac{\alpha}{\theta})\phi(\bar{i}) - \gamma - \delta\right]\frac{1}{\gamma + \delta}(\bar{i} - q(\bar{i})) = a\phi(\bar{i})^{1-\frac{1}{\alpha}} + q(\bar{i}). \tag{7.38}$$

Equation (7.37) is equivalent to (7.10), with b given by (7.17).

7.6.2 The Proof of Proposition 6

Within our ranges $(0, i_1)$ and $(i_2, +\infty)$, $\phi(i)$ is positive, and hence $f(i)$ is continuous and differentiable. In particular,

$$f'(i) = \phi'(i)\left\{b[i - q(i)] + a(\frac{1}{\alpha} - 1)\phi(i)^{-\frac{1}{\alpha}}\right\} + (bm - 1), \tag{7.39}$$

where

$$\phi'(i) = \frac{mq''(i)}{\left[1 - q'(i)\right]^2}. \tag{7.40}$$

First, a, b, and m are all positive, as indicated by (7.16), (7.17), and (7.19), and therefore the term $b[i - q(i)] + a(\frac{1}{\alpha} - 1)\phi(i)^{-\frac{1}{\alpha}}$ is positive. In the range $(0, i_1)$, $q''(i) > 0$, and hence $f'(i) > 0$. However, in the range $(i_2, +\infty)$, $q''(i) < 0$, and hence $f'(i)$ can be either positive or negative, according to the sign of $(bm - 1)$.

Let us first consider the range $(0, i_1)$. Assume $i \to 0$. In this case, $q(i) \to 0$, and therefore $f(i) \to -a\phi(0)^{1-\frac{1}{\alpha}} < 0$. Next, assume $i \to i_1$. In this case, $q'(i) \to 1$, and therefore $\phi(i) \to +\infty$. Since $1 - \frac{1}{\alpha}$ is negative, this further indicates that $\phi(i)^{1-\frac{1}{\alpha}} \to 0$. Therefore, $f(i) \to +\infty$. Since $f'(i) > 0$, by the intermediate value theorem there exists one and only one i such that $f(i) = 0$. We thus have proved the first part of the proposition.

Next we turn to the range $(i_2, +\infty)$. To verify the second part of the proposition, we only need to prove that $f(i) \to +\infty$ and $f(i) < 0$ when $i \to i_2$

and $f(i) \to +\infty$, and $f(i) > 0$ when $i \to +\infty$. Consider first $i \to i_2$. Again, $q'(i) \to 1$, $\phi(i) \to +\infty$, and $\phi(i)^{1-\frac{1}{\alpha}} \to 0$. Therefore $f(i) \to +\infty$. Meanwhile, from (7.39), $\phi'(i) \to -\infty$ (since $q''(i) < 0$), and therefore $f'(i) < 0$. Consider now $i \to +\infty$. In this case, $q'(i) \to 0$ and $q(i) \to q^m$, where q^m is the upper limit of $q(i)$. This indicates that $[i - q(i)] \to +\infty$. Therefore, $f(i) \to +\infty$. Meanwhile, since $q''(i) \to 0$, $\phi'(i) \to 0$. Therefore $f'(i) \to (bm - 1)$, which is positive. We thus have proved the second part of the proposition.

8

Business Cycles with a Nonclearing Labor Market

8.1 Introduction

As discussed in the previous chapters, especially in chapter 5, the standard real business cycle (RBC) model, despite its rather simple structure, can explain the volatilities of macroeconomic variables such as output, consumption, and capital stock. However, to explain the actual variation in employment, the model generally predicts an excessive smoothness of labor effort in contrast to empirical data. This problem of excessive smoothness in labor effort and some incorrect correlations of labor effort with other macrovariables are well known in the RBC literature. An evaluation of this failure of the RBC model is given in Schmidt-Grohe (2001).[1] There the RBC model is also compared with indeterminacy models, as developed by Benhabib and his coauthors. Both types of models presume flexible wages and prices. Whereas in RBC models the standard deviation of the labor effort is too low, in indeterminacy models it turns out to be excessively high. It has preliminarily been explored in chapter 5 of this volume. Last, the RBC model predicts a significantly high positive correlation between technology and employment, whereas empirical research demonstrates, at least at business cycle frequency, a negative or almost 0 correlation.

These are the major issues that we take up from now on. The labor market problems, the lack of variation in the employment and the correlation between consumption and macrovariables in the standard RBC model may be related to the specification of the labor market, and therefore we name it the labor market puzzle. In this chapter we are mainly concerned with this puzzle. The technology puzzle—the excessively high correlation between technology and employment, preliminarily discussed in chapter 5, will be taken up in chapter 9.

Although in the specification of its structure (see chapter 4) the RBC model presents both demand and supply of a market, the moments of the economy are reflected by the variation on one side of the market due to its general equilibrium for all markets (including output, labor, and capital markets). For the labor market, the moments of labor effort result from the decision rule of the representative household to supply labor, yet the household always equates the marginal disutility of work to the utility of the real wage and the firm equates the marginal product of labor to the real wage. In the approximate solution of the decision variables in the standard RBC model, the variations in labor effort and consumption reflect the moments of the two state variables, capital and technology. Since both are choice variables in the RBC model and respond to the two state variables it is therefore not surprising that the variation of consumption is as smooth as labor effort. This further suggests that to resolve the labor market puzzle, one has to improve labor market specifications. One possible approach for such improvement is to introduce Keynesian features into the model by allowing for monopolistic competition, price and wage stickiness, and a nonclearing labor market.

Historically, research along the line of microbased Keynesian economics has been developed by two approaches: disequilibrium analysis, which was popular before the 1980s, and the New Keynesian analysis based on monopolistic competition. There have been recent attempts to introduce the Keynesian features into a dynamic optimization model. Rotemberg and Woodford (1995, 1999), King and Wolman (1999), Gali (1999), and Woodford (2003) present models with monopolistic competition and sticky prices. On the other hand, there are models of efficiency wages in which a nonclearing labor market could occur.[2] In those studies with a nonclearing labor market, an explicit labor demand function is introduced from the decision problem of the firm side. However, the decision rule with regard to labor supply in these models is often dropped because the labor supply no longer appears in the utility function of the household. Consequently, the moments of labor effort become purely demand determined.[3]

In this chapter, we present a stochastic dynamic optimization model of the RBC type augmented by Keynesian features along the line of the above considerations. In particular, we allow for price and wage stickiness[4] and a nonclearing labor market. However, unlike other recent models that drop the decision rule of labor supply, we view the decision rule of the labor effort as being derived from a dynamic optimization problem as a quite natural way to determine desired labor supply.[5] With the determination of labor demand, derived from the marginal product of labor and other factors,[6] the two basic forces in the labor market can be formalized. One of the advantages of this formulation, as will become clear, is that a variety of employment rules can be adopted to specify the realization of actual employment when a nonclearing market emerges.[7] We assess this model by employing U.S. and German macroeconomic time series data. An important feature of our new model is, as maintained in the Keynesian literature, that consumption is linked not only to the future path of the state variables but also to current employment.[8]

Yet before we formally present the model and its calibration, we will note that there is a similarity of our approach and the New Keynesian analysis. New Keynesian literature presents models with imperfect competition and sluggish price and wage adjustments where labor effort is endogenized. Important work of this type can be found in Rotemberg and Woodford (1995, 1999), King and Wolman (1999), Gali (1999), Erceg et al. (2000), and Woodford (2003). However, in those models the markets are still assumed to be cleared because the producer supplies the output according to what the market demands at the existing price. A similar consideration is assumed to hold for the labor market. Here the wage rate is set optimally by a representative of the household according to the expected market demand curve for labor. Once the wage has been set, it is assumed to be sticky for some time period, and only a fraction of wages are set optimally in each period. In those models there will be a gap between the optimal wage and the existing wage, yet the labor market is cleared because the household is assumed to supply labor, whatever the market demand is, at the given wage rate.[9]

In this chapter, we present a dynamic model that allows for a noncleared labor market, which could be seen as being caused by staggered wages, as described by Taylor (1980) and Calvo (1983), or other theories of sluggish wage adjustment. The objective in constructing a model like ours is to approach the two aforementioned labor market problems coherently within a single model of dynamic optimization. Yet, we wish to argue that the New Keynesian and disequilibrium approaches are complementary rather than mutually exclusive, and therefore they can be somewhat consolidated as a more complete system for price and quantity determination within the Keynesian tradition. (For further details of this consolidation, see chapter 9.) In the present chapter we are concerned only with a nonclearing of the labor market as brought into the academic discussion by the disequilibrium school. We derive the nonclearing of the labor market from optimizing behavior of economic agents, but it will be a multiple stage decision process that will generate the nonclearing of the labor market.[10]

The remainder of this chapter is organized as follows. Section 8.2 presents the model structure. Section 8.3 estimates and calibrates the model variants for the U.S. economy. Section 8.4 undertakes the same exercise for the German economy. Section 8.5 concludes. Appendices I and II contain technical derivations of the adaptive optimization procedure, and appendix III undertakes a welfare comparison of the model variants.

8.2 An Economy with a Nonclearing Labor Market

We will still follow the usual assumptions of identical households and identical firms. Therefore we are considering an economy that has two representative agents: the representative household and the representative firm. There are three markets in which the agents exchange their products, labor, and capital. The household owns all the factors of production and therefore sells factor services to the firm. The revenue from selling factor services can be used only to buy the goods produced by the firm, either for consuming or for accumulating capital. The representative firm owns nothing. It simply hires capital and labor to produce output, sells the output, and transfers the profit back to the household.

Unlike the typical RBC model, in which one could assume a once-for-all market, in this model we will assume that the market is to be reopened at the beginning of each period t. This is necessary for a model with nonclearing markets in which adjustments should occur that lead us to a multiple-stage adaptive optimization behavior. Let us first describe how prices and wages are set.

8.2.1 The Wage Determination

As usual, we presume that the household and the firm express their desired demand and supply on the basis of given prices, including the output price p_t, the wage rate w_t, and the rental rate of capital stock r_t. We first discuss how the period t prices are determined at the beginning of period t. There are three commodities in our model. One of them should serve as a numeraire, which we assume to be the output. Therefore, the output price p_t always equals 1. This indicates that the wage w_t and the rental rate of capital stock r_t are measured in terms of the physical units of output.[11] The rental rate of capital r_t is assumed to be adjustable so as to clear the capital market. We can then ignore its setting. Indeed, as will become clear, one can imagine any initial value of the rental rate of capital when the firm and the household make the quantity decisions and express their desired demand and supply. This leaves us to focus the discussion on the wage setting. Let us first discuss how the wage rate might be set.

Most recent literature, in discussing wage setting,[12] assumes that it is the supplier of labor—the household or its representative—which sets the wage rate; the firm is simply a wage taker. On the other hand, there are models that discuss how firms set the wage rate.[13] In actual bargaining it is likely, as Taylor (1999) has pointed out, that wage setting is an interactive process between firms and households. Despite this variety of wage-setting models, we follow the recent approach. We may assume that the wage rate is set by a representative of the household who acts as a monopolistic agent for the supply of labor effort, as Woodford (2003, chapter 3) has suggested. Woodford (p. 221) introduces different wage-setting agents and monopolistic competition because he assumes heterogeneous households are different suppliers of differentiated types of labor. In appendix I, in close relationship to Woodford (chapter 3), Erceg et al. (2000), and Christiano et al. (2001), we present a wage-setting model in which wages are set optimally, but a fraction of wages may be sticky. We neglect differentiated types of labor and refer only to aggregate wages.

However, recently many theories have been developed to explain wage and price stickiness. There is the so-called menu cost for changing prices (though this seems more appropriate for the output price). There is also a reputation cost for changing prices and wages.[14] In addition, changing the price, or wage, requires information, computation, and communication, which may be costly.[15] All these efforts involve costs that may be summarized as adjustment costs of changing the price or wage. The adjustment cost for changing the wage may provide some reason for the representative of the household to stick to the wage rate even if it is known that the current wage may not be optimal. One may also derive this stickiness of wages from wage contracts, as in Taylor (1980), with the contract period to be longer than one period.

Since workers, or their representative, usually enter into long-term employment contracts involving labor supply for several periods, with a variety of job security arrangements and termination options, a wage contract may also be understood from an asset price perspective: as derivative security based on a fundamental underlying asset, such as the asset price of the firm. In principle, a wage contract could be treated as a debt contract with long-term commitments similar to those for other liabilities of the firm.[16] As in the case of pricing of corporate liabilities, the wage contract—the value of the derivative security—depends on specifications in contractual agreements. In general, however, it can be assumed to be arranged for several periods.

As noted above, we do not have to posit that the wage rate, w_t, is completely fixed in contracts and never responds to the disequilibrium in the labor market. One may imagine that the dynamics of the wage rate, for example, follows the updating scheme suggested in Calvo's staggered price model (1983) or in Taylor's wage contract model (1980). In Calvo's model, for example, there is always a fraction of individual prices to be adjusted in each period t.[17] This can be expressed in our model as the expiration of some wage contracts, and therefore new wage contracts will be signed in each t. The new signed wage contracts should respond to the expected market conditions not only in period t but also through to $t + j$, where j can be regarded as the contract period.[18] Through such a pattern of wage dynamics, wages are only partially adjusted.

Explicit formulation of wage dynamics through a Calvo type of updating scheme, particularly with differentiated types of labor, is studied in Erceg et al. (2000), Christiano et al. (2001), and Woodford (2003, chapter 3), and briefly sketched, as underlying our model, for an aggregate wage in appendix I of this chapter. A more explicit treatment is not needed here. Indeed, as will become clear in section 8.3, the empirical study of our model does not rely on how we formulate the wage dynamics. All we need to presume is that wage contracts are only partially adjusted, giving rise to a sticky aggregate wage.[19]

8.2.2 The Household's Desired Transactions

The next step in our multiple-stage decision process is to model the quantity decisions of the households. When the price, including the wage, has been set, the household will express its demand for goods and its supply of factors. We define the household's desired demand and supply as those which allow the household to obtain the maximum utility, on the condition that this demand and this supply can be realized at the given set of prices. We can express the household's desired demand and supply as a sequence of output demands and factor supplies $\left\{c_{t+i}^d, i_{t+i}^d, n_{t+i}^s, k_{t+i+1}^s\right\}_{i=0}^{\infty}$, where i_{t+i} refers to investment. Here we have used the superscripts d and s to refer to the agent's desired demand and supply, respectively. The decision problem for the household to derive its demand and supply can be formulated as

$$\max_{\{c_{t+i}, n_{t+i}\}_{i=0}^{\infty}} E_t \left[\sum_{i=0}^{\infty} \beta^i U(c_{t+i}^d, n_{t+i}^s) \right], \tag{8.1}$$

subject to

$$c_{t+i}^d + i_{t+i}^d = r_{t+i} k_{t+i}^s + w_{t+i} n_{t+i}^s + \pi_{t+i} \tag{8.2}$$

$$k_{t+i+1}^s = (1 - \delta) k_{t+i}^s + i_{t+i}^d. \tag{8.3}$$

Above, π_{t+i} is the expected dividend. Equation (8.2) can be regarded as a budget constraint. The equality holds because of the assumption $U_c > 0$. Next, we consider how the representative household calculates π_{t+i}.

Assuming that the household knows the production function $f(\cdot)$ while it expects that all its optimal plans can be fulfilled at the given price sequence $\left\{p_{t+i}, w_{t+i}, r_{t+i}\right\}_{i=0}^{\infty}$, we obtain

$$\pi_{t+i} = f\left(k_{t+i}^s, n_{t+i}^s, A_{t+i}\right) - w_{t+i} n_{t+i}^s - r_{t+i} k_{t+i}^s. \tag{8.4}$$

Explaining π_{t+i} in (8.2) in terms of (8.4) and then substituting from (8.3) to eliminate i_t^d, we obtain

$$k_{t+i+1}^s = (1 - \delta) k_{t+i}^s + f\left(k_{t+i}^s, n_{t+i}^s, A_{t+i}\right) - c_{t+i}^d. \tag{8.5}$$

For the given technology sequence $\{A_{t+i}\}_{i=0}^{\infty}$, equations (8.1) and (8.5) form a standard intertemporal decision problem. The solution to this problem can be written as

$$c_{t+i}^d = G_c\left(k_{t+i}^s, A_{t+i}\right) \tag{8.6}$$

$$n_{t+i}^s = G_n\left(k_{t+i}^s, A_{t+i}\right). \tag{8.7}$$

Although the solution appears to be a sequence $\left\{c_{t+i}^d, n_{t+i}^s\right\}_{i=0}^{\infty}$, only (c_t^d, n_t^s) and (i_t^d, k_t^s), where $i_t^d = f(k_t^s, n_t^s, A_t) - c_t^d$ and $k_t^s = k_t$, are actually carried into the market by the household for exchange because of our assumption that the market reopens.

8.2.3 The Firm's Desired Transactions

As in the case of the household, the firm's demand for factors and its supply of goods are those which maximize the firm's profit under the condition that all its goals can be attained at the given set of prices. The optimization problem for the firm thus can be expressed as being to choose the input demands and output supplies (n_t^d, k_t^d, y_t^s) that maximize the current profit:

$$\max\ y_t^s - r_t k_t^d - w_t n_t^d,$$

subject to

$$y_t^s = f\left(A_t, k_t^d, n_t^d\right). \tag{8.8}$$

For regular conditions on the production function $f(\cdot)$, the solution to the above optimization problem should satisfy

$$r_t = f_k\left(k_t^d, n_t^d, A_t\right) \tag{8.9}$$

$$w_t = f_n\left(k_t^d, n_t^d, A_t\right), \tag{8.10}$$

where $f_k(\cdot)$ and $f_n(\cdot)$ are the marginal products of capital and labor, respectively. Next we consider the transactions in our three markets. Let us first consider the two factor markets.

8.2.4 Transactions in the Factor Market and Actual Employment

We have assumed the rental rate of capital r_t to be adjustable in each period, and thus the capital market is cleared. This indicates that

$$k_t = k_t^s = k_t^d.$$

Concerning the labor market, there is no reason to believe that the firm's demand for labor, as implicitly expressed in (8.9), should be equal to the willingness of the household to supply labor as determined in (8.7), given the way wage determination is explained in section 8.2.1. Therefore, we cannot regard the labor market to be cleared. An illustration of this statement, in a simpler version, is given in appendix I.[20]

Given a nonclearing labor market, we have to specify what rule should apply regarding the realization of actual employment.

Disequilibrium Rule: When disequilibrium occurs in the labor market, either of the following rules will be applied:

$$n_t = \min\left(n_t^d, n_t^s\right) \tag{8.11}$$

$$n_t = \omega n_t^d + (1 - \omega) n_t^s, \tag{8.12}$$

where $\omega \in (0, 1)$.

Above, (8.11) is the famous short-side rule when non-clearing of the market occurs. It has been widely used in the literature on disequilibrium analysis (see, e.g., Bénassy 1975, 1984). And (8.12) might be called the compromising rule. This rule indicates that when non-clearing of the labor market occurs, both firms and workers have to compromise. If there is excess supply, firms will employ more labor than they wish to employ.[21] On the other hand, when there is excess demand, workers will have to offer more effort than they wish to offer.[22] Such compromises may be the result of institutional structures and moral standards of the society.[23] Given the rather corporate relationship of labor and firms in Germany, for example, this compromising rule might be considered a reasonable approximation. Such a rule that seems to hold for many other countries was discussed early in the economic literature (see Meyers 1968; Solow 1979).

The unemployment we discuss here is different from the frictional unemployment often discussed in search and matching models. In our representative agent model, the unemployment is mainly due to adaptive optimization of the household, given the institutional arrangements of setting wages (see section 8.5). Frictional unemployment can arise from informational and institutional search and matching frictions in which welfare state and labor market institutions may play a role.[24] Yet the frictions in the institutions of the matching process are likely to explain only a fraction of observed unemployment.[25]

8.2.5 Actual Employment and Transactions in the Product Market

After the transactions in these two factor markets have been carried out, the firm will engage in production activity. The result is the output supply, which, instead of (8.8), is now given by

$$y_t^s = f(k_t, n_t, A_t). \tag{8.13}$$

Then the transaction needs to be carried out with respect to y_t^s. It is important to note that when the labor market is not cleared, the consumption plan expressed by (8.6) becomes invalid because of the improper budget constraint (8.2)—which further brings the improper transition law of capital (8.5)—for deriving the plan. Therefore, the household will be required to construct a new consumption plan, which should be derived from the following optimization program:

$$\max_{(c_t^d)} U(c_t^d, n_t) + E_t \left[\sum_{i=1}^{\infty} \beta^i U\left(c_{t+i}^d, n_{t+i}^s\right) \right], \tag{8.14}$$

subject to

$$k_{t+1}^s = (1-\delta)k_t + f(k_t, n_t, A_t) - c_t^d \tag{8.15}$$

$$k_{t+i+1}^s = (1-\delta)k_{t+i}^s + f\left(k_{t+i}^s, n_{t+i}^s, A_{t+i}\right) - c_{t+i}^d \tag{8.16}$$

$$i = 1, 2, \ldots$$

Note that in this optimization program the only decision variable is about c_t^d, and the data include not only A_t and k_t but also n_t, which is given by either (8.11) or (8.12). We can write the solution in terms of the following equation (see appendix II of this chapter for details):[26]

$$c_t^d = G_{c2}(k_t, A_t, n_t). \tag{8.17}$$

Given this adjusted consumption plan, the product market should be cleared if the household demands the amount $f(k_t, n_t, A_t) - c_t^d$ for investment. Therefore, c_t^d in (8.17) should also be the realized consumption.

8.3 Estimation and Calibration for the U.S. Economy

This section provides an empirical study, for the U.S. economy, of our model as presented in section 8.2. However, the model in 8.2 is only for illustrative purposes. It is not the model that can be tested with empirical data, not only because we do not specify the forms of production function, utility function, and the stochastic process of A_t, but also because we do not introduce the growth factor into the model. For an empirically testable model, we here employ the model formulated by King et al. (1988a,b).

8.3.1 The Empirically Testable Model

Let K_t denote capital stock; N_t, per capita working hours; Y_t, output; and C_t, consumption. Assume that the capital stock in the economy follows the transition law:

$$K_{t+1} = (1-\delta)K_t + A_t K_t^{1-\alpha}(N_t X_t)^{\alpha} - C_t, \tag{8.18}$$

where δ is the depreciation rate; α is the share of labor in the production function $F(\cdot) = A_t K_t^{1-\alpha}(N_t X_t)^{\alpha}$; A_t is the temporary shock in technology; and X_t is the permanent shock that follows a growth rate γ.[27] The model is nonstationary due to X_t. To transform the model into a stationary setting, we divide both sides of equation (8.18) by X_t:

$$k_{t+1} = \frac{1}{1+\gamma}\left[(1-\delta)k_t + A_t k_t^{1-\alpha}(n_t \bar{N}/0.3)^{\alpha} - c_t\right], \tag{8.19}$$

where $k_t \equiv K_t/X_t$, $c_t \equiv C_t/X_t$, and $n_t \equiv 0.3N_t/\bar{N}$, with $\bar{N}$ the sample mean of N_t. Note that n_t is often regarded as the normalized hours. The sample mean of n_t is equal to 30%, which, as pointed out by Hansen (1985), is the average percentage of hours attributed to work. The above formulation also indicates that the form of $f(\cdot)$ in section 8.2 may follow

$$f(\cdot) = A_t k_t^{1-\alpha}(n_t \bar{N}/0.3)^{\alpha}, \tag{8.20}$$

while $y_t \equiv Y_t/X_t$ with Y_t the empirical output.

With regard to the household preference, we assume that the utility function takes the form

$$U(c_t, n_t) = \log c_t + \theta \log (1-n_t). \tag{8.21}$$

The temporary shock A_t may follow an AR(1) process:

$$A_{t+1} = a_0 + a_1 A_t + \epsilon_{t+1}, \tag{8.22}$$

where ϵ_t is an independently and identically distributed (i.i.d.) innovation: $\epsilon_t \sim N(0, \sigma_\epsilon^2)$.

8.3.2 The Data-Generating Process

For our empirical test, we consider three model variants: the standard RBC model (as a benchmark for comparison) and the two labor market disequilibrium models with the disequilibrium rules as expressed in (8.11) and (8.12). Specifically, we will call the standard model, model I; the disequilibrium model with the short-side rule (8.11), model II; and the disequilibrium model with the compromising rule (8.12), model III.

For the standard RBC model, the data-generating process includes (8.19), (8.22), and

$$c_t = G_{11}A_t + G_{12}k_t + g_1 \tag{8.23}$$

$$n_t = G_{21}A_t + G_{22}k_t + g_2. \tag{8.24}$$

Here (8.23) and (8.24) are the linear approximations to (8.6) and (8.7) when we ignore the superscripts s and d. The coefficients G_{ij} and $g_i (i = 1, 2$ and $j = 1, 2)$ are the complicated functions of the model's structural parameters, α and β, among others. They are computed, as in chapter 5, by the numerical algorithm using the linear-quadratic approximation method presented in chapters 1 and 2. For our chosen parameters (see table 8.1) we numerically obtain

$$G_{11} = 172.864, \quad G_{12} = 0.9634$$
$$G_{21} = 0.00356, \quad G_{22} = -0.00012.$$

Given these coefficients and the parameters in equation (8.22), including σ_ε (also reported in table 8.1), we can simulate the model to generate stochastically simulated data. These data can then be compared with the sample moments of the observed economy.

Obviously, the standard model does not allow for non-clearing of the labor market. The moments of the labor effort are reflected solely by the decision rule (8.24), which is quite similar in its structure to the decision rule given by (8.23)—that is, they are both determined by k_t and A_t. Since the technology shock is dominant and the coefficient G_{22} is only very slightly negative,[28] the structural similarity of the linear decision rules (8.23) and (8.24) is expected to produce two labor market problems: the volatility of the labor effort cannot be much different from the volatility of consumption, which generally appears to be too smooth, and some cross-correlations of labor effort with macrovariables do not fit the data. This holds for our parameter constellations.

To define the data-generating process for our disequilibrium models, we first modify (8.24) as

$$n_t^s = G_{21}A_t + G_{22}k_t + g_2. \tag{8.25}$$

On the other hand, the equilibrium in the product market indicates that c_t^d in (8.17) should be equal to c_t. As derived in appendix II, this equation can be approximated as

$$c_t = G_{31}A_t + G_{32}k_t + G_{33}n_t + g_3. \tag{8.26}$$

In appendix II, we provide the details of how to compute the coefficients $G_{3j}, j = 1, 2, 3$, and g_3. We observe here that in the nonclearing labor market model, given the constraint that the household faces on the labor market, the consumption path is impacted by actual employment, n_t.[29]

Next we consider the labor demand derived from the production function $F(\cdot) = A_t K_t^{1-\alpha}(N_t X_t)^\alpha$. Let $X_t = Z_t L_t$, with Z_t the permanent shock resulting purely from productivity growth, and L_t, the shock from population growth. We assume that L_t has a constant growth rate μ, and hence Z_t follows the growth rate $(\gamma - \mu)$. The production function can be written as $Y_t = A_t Z_t^{\alpha} K_t^{1-\alpha} H_t^{\alpha}$, where H_t equals $N_t L_t$, which can be regarded as total labor hours. Taking the partial derivative with respect to H_t and recognizing that the marginal product of labor is equal to the real wage, we obtain

$$w_t = \alpha A_t Z_t k_t^{1-\alpha} (n_t^d \bar{N}/0.3)^{\alpha-1}.$$

This equation is equivalent to (8.10). It generates the demand for labor as

$$n_t^d = (\alpha A_t Z_t / w_t)^{1/(1-\alpha)} k_t (0.3/\bar{N}). \tag{8.27}$$

The per capita hours demanded, n_t^d, should be stationary if the real wage w_t and productivity Z_t grow at the same rate. This seems to be roughly consistent with the U.S. experience that we now calibrate.

Thus, for the nonclearing market model with the short-side rule, model II, the data-generating process includes (8.19), (8.22), (8.11), (8.25), (8.26), and (8.27), with w_t given by the observed wage rate. We do not attempt here to give the actually observed sequence of wages a further theoretical foundation.[30] For our purpose it suffices to take the empirically observed series of wages. For model III, we use (8.12) instead of (8.11).

8.3.3 The Data and the Parameters

Before we calibrate the models, we will specify the parameters. Altogether there are ten parameters in our three variants: a_0, a_1, σ_ε, γ, μ, α, β, δ, θ, and ω. We first specify α and γ, respectively, at 0.58 and 0.0045, which are standard. This allows us to compute the data series of the temporary shock A_t. With this data series, we estimate the parameters a_0, a_1, and σ_ε. The next three parameters—β, δ, and θ—are estimated with the GMM method by matching the moments of the model generated by (8.19), (8.23), and (8.25). The estimation is conducted by a global optimization algorithm called simulated annealing. These parameters have been estimated in chapter 5, and therefore we employ them here. For the new parameters, we specify μ at 0.001, which is close to the average growth rate of the labor force in the United States; the parameter ω in model III is set to 0.1203. It is estimated by minimizing the residual sum of the squares of actual employment and the model-generated employment. The estimation is executed by a conventional algorithm, the grid search. Table 8.1 presents these parameters.

The data set used in this section is taken from Christiano (1987). The wage series was obtained from Citibase. It is rescaled to match the model's implication.[31]

Table 8.1. Parameters Used for Calibration

a_0	0.0333	σ_ε	0.0185	μ	0.0010	β	0.9930	θ	2.0189
a_1	0.9811	γ	0.0045	α	0.5800	δ	0.2080	ω	0.1203

8.3.4 Calibration

Table 8.2 reports our calibration from 5,000 stochastic simulations. The results in this table are confirmed by figure 8.1, where a one-time simulation with the observed innovation A_t is presented.[32] All time series are detrended by the HP-filter.

First, the structural parameters that we used here for calibration are estimated by matching the model I economy to the sample economy. The result, reflected in table 8.2, is somewhat biased in favor of the model I economy. It is not surprising that for most variables the moments generated from the model I economy are closer to the moments of the sample economy. Yet even in this case, there is an excessive smoothness of the labor effort and the employment series of the data cannot be matched. For our time period, 1955.1 to 1983.4, in the model I economy we find 0.32 as the ratio of the standard deviation of labor effort to the standard deviation of output. This ratio is roughly 1 in the sample economy. The problem is, however, resolved in model II and model III economies, representing sticky wages and labor market non-clearing. There the ratios are 1.38 and 0.69 for model II and model III economies, respectively.

Further evidence on the better fit of the nonclearing labor market models—concerning the volatility of the macroeconomic variables—is demonstrated in figure 8.1, where the graphs show actual data (solid line) and simulated data (dotted line) for consumption, capital stock, employment, and output, the three columns representing the figures for model I, model II and model III economies. As can be seen, the model III economy in particular best fits, along most dimensions, the actual data. The volatility of employment greatly increases for models II and III. In particular, the volatility of the model III economy is close to that of the sample economy. An excessively high volatility is observable in the model II economy, which may reflect our assumption that there are no search and matching frictions (which, of course, in the actual economy will not hold). We therefore may conclude that model III best matches the labor market volatility.

The failure of the standard model to match the volatility of employment data is described by Schmidt-Grohe (2001). For her employment time series (1948.3–1997.4), she finds that the ratio of the standard deviation of employment to the standard deviation of output is roughly 0.95, close to our sample economy. Yet for the standard RBC model, the ratio is 0.49, which is too low compared with the empirical data. For the indeterminacy model,

Table 8.2. Calibration of the Model Variants: U.S. Economy

	Consumption	Capital	Employment	Output
Standard Deviation				
Sample Economy	0.0081	0.0035	0.0165	0.0156
Model I Economy	0.0091	0.0036	0.0051	0.0158
	(0.0012)	(0.0007)	(0.0006)	(0.0021)
Model II Economy	0.0137	0.0095	0.0545	0.0393
	(0.0098)	(0.0031)	(0.0198)	(0.0115)
Model III Economy	0.0066	0.0052	0.0135	0.0197
	(0.0010)	(0.0010)	(0.0020)	(0.0026)
Correlation Coefficient				
Sample Economy				
Consumption	1.0000			
Capital Stock	0.1741	1.0000		
Employment	0.4604	0.2861	1.0000	
Output	0.7550	0.0954	0.7263	1.0000
Model I Economy				
Consumption	1.0000			
	(0.0000)			
Capital Stock	0.2043	1.0000		
	(0.1190)	(0.0000)		
Employment	0.9288	−0.1593	1.0000	
	(0.0203)	(0.0906)	(0.0000)	
Output	0.9866	0.0566	0.9754	1.0000
	(0.00332)	(0.1044)	(0.0076)	(0.0000)
Model II Economy				
Consumption	1.0000			
	(0.0000)			
Capital Stock	0.4944	1.0000		
	(0.1662)	(0.0000)		
Employment	0.4874	−0.0577	1.0000	
	(0.1362)	(0.0825)	(0.0000)	
Output	0.6869	0.0336	0.9392	1.0000
	(0.1069)	(0.0717)	(0.0407)	(0.0000)
Model III Economy				
Consumption	1.0000			
	(0.0000)			
Capital Stock	0.4525	1.0000		
	(0.1175)	(0.0000)		
Employment	0.6807	−0.0863	1.0000	
	(0.0824)	(0.1045)	(0.0000)	
Output	0.8924	0.0576	0.9056	1.0000
	(0.0268)	(0.0971)	(0.0327)	(0.0000)

Note: Numbers in parentheses are the corresponding standard errors.

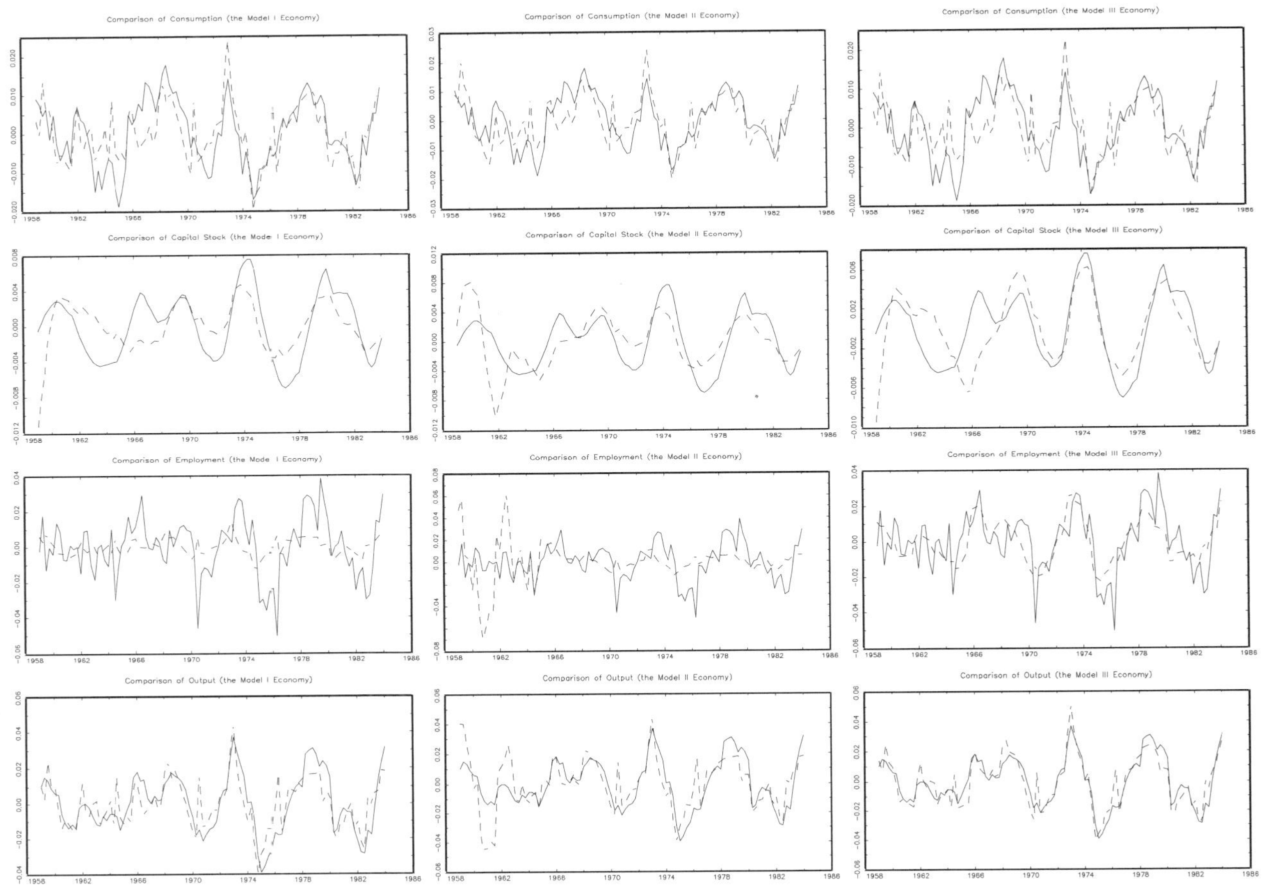

Figure 8.1. Simulated Economy versus U.S. Economy. Note: Solid line is U.S. economy; dashed line is simulated economy.

originating in the work by Benhabib et al., the ratio is 1.45, which seems too high. As noted above, a similarly high ratio of standard deviations is also observed in our model II economy, where the short-side rule leads to excessive fluctuations of the labor effort.

Next, let us look at the cross-correlations of the macroeconomic variables. In the sample economy, there are two significant correlations: between consumption and output (roughly 0.75) and between employment and output (about 0.72). These two strong correlations are also found in all of our simulated economies. Yet, the appropriate correlation between consumption and employment is replicated only in our model II and model III economies. Matching the empirical correlation between consumption and employment has recently become an important issue,[33] and our model II and model III economies are able to do so.

A success of our nonclearing labor market models (see the model II and model III economies) is that employment is no longer significantly correlated with consumption. This is because we have made a distinction between the demand for and the supply of labor, whereas only labor supply reflects the moments of capital and technology as consumption does. Since the realized employment is not necessarily the same as the labor supply, the correlation with consumption is therefore weakened.

8.4 Estimation and Calibration for the German Economy

Above we employed a model with nonclearing labor market for the U.S. economy. We saw that one of the major reasons the standard model cannot appropriately replicate the variation in employment is its not introducing the demand for labor. Next, we pursue a similar study of the German economy. We will first summarize some stylized facts of the German economy compared with the U.S. economy.

8.4.1 The Data

Our study of the German economy employs time series data from 1960.1 to 1992.1. We thus have included a short period after the unification of Germany (1990–1991). We again use quarterly data. The time series data on GDP, consumption, investment, and capital stock are from the OECD (see OECD 1998a), and the data on total labor force are also from the OECD (1998b). The time series data on total working hours are from Statistisches Bundesamt (1998). The time series on the hourly real wage index is from the OECD (1998a).

8.4.2 The Stylized Facts

Next, we want to compare some stylized facts. Figures 8.2 and 8.3 compare six key variables relevant for the models for the German and U.S. economies.

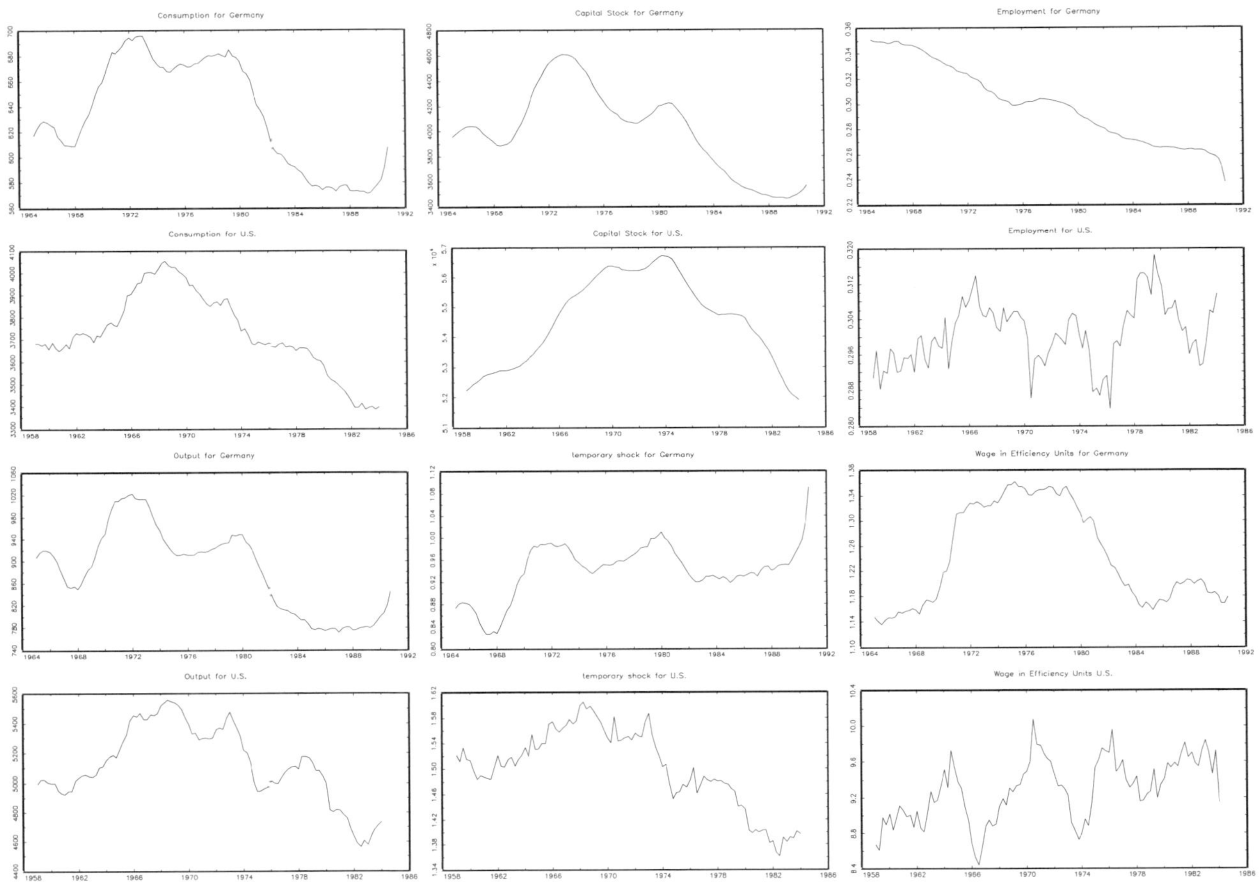

Figure 8.2. Comparison of Macroeconomic Variables, United States and Germany

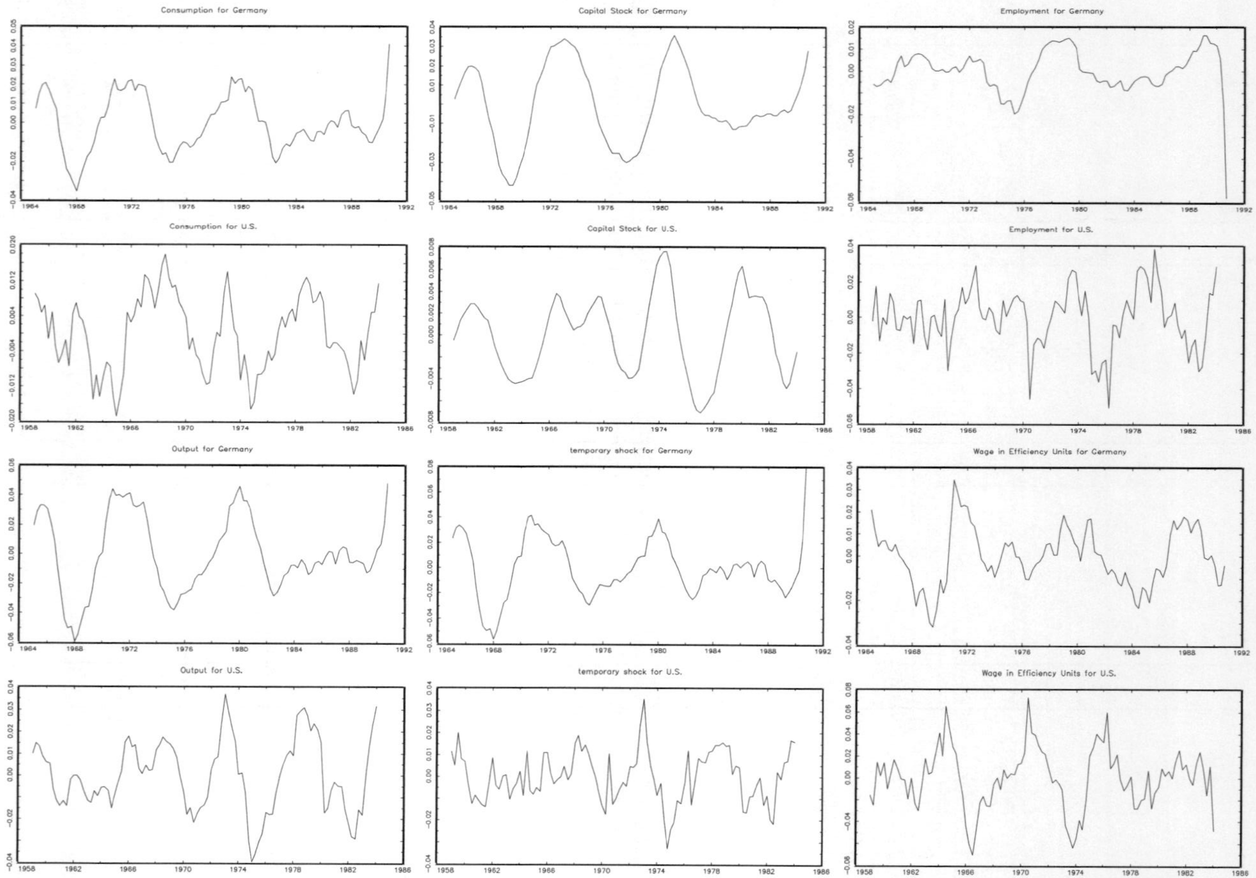

Figure 8.3. Comparison of Macroeconomic Variables, United States and Germany: Data Series Detrended by HP-Filter

Table 8.3. The Standard Deviations: United States and Germany

	Germany (detrended)	U.S. (detrended)
Consumption	0.0146	0.0084
Capital Stock	0.0203	0.0036
Employment	0.0100	0.0166
Output	0.0258	0.0164
Temporary Shock	0.0230	0.0115
Efficiency Wage	0.0129	0.0273

The data in figure 8.3 are detrended by the HP-filter. The standard deviations of the detrended series are summarized in table 8.3.

Several remarks are in order here. First, employment and the efficiency wage are among the variables with the highest volatility in the U.S. economy. However, in the German economy they are the smoothest variables. Second, employment (measured in terms of per capita hours) is declining over time in Germany (see figure 8.2 for the nondetrended series), while in the U.S. economy, the series is approximately stationary. Third, in the U.S. economy, the capital stock and temporary shock to technology are both relatively smooth. In contrast, they are both more volatile in Germany. These results might be due to the difference in employment volatility. The volatility of output must be absorbed by factors in the production function. If employment is smooth, the other two factors must be volatile.

Should we expect that such differences will lead to different calibrations of our model variants? This will be explored next.

8.4.3 The Parameters

For the German economy, our investigation showed that an AR(1) process does not match the observed process of A_t well. Instead, we will use an AR(2) process:

$$A_{t+1} = a_0 + a_1 A_t + a_2 A_{t-1} + \varepsilon_{t+1}.$$

The parameters used for calibration are given in table 8.4. All of these parameters are estimated in the same way as those for the U.S. economy.

Table 8.4. Parameters Used for Calibration: German Economy

a_0	0.0044	γ	0.0083	δ	0.0538
a_1	1.8880	μ	0.0019	θ	2.1507
a_2	−0.8920	α	0.6600	ω	0
σ_ε	0.0071	β	0.9876		

It is important to note that the estimated ω in this case is on the boundary 0, indicating that the weight of the demand is 0[34] in the compromising rule (8.12). In other words, the model III economy is almost identical to the model I economy. This seems to indicate that the model I economy, the standard model, will be the best in matching German labor market.

8.4.4 Calibration

In table 8.5 we provide the calibration results from 5,000 stochastic simulations for the German economy. In figure 8.4 we compare the one-time simulation with the observed A_t for our model variants. Again all time series are detrended by the HP-filter.[35]

Table 8.5. Calibration of the Model Variants: German Economy

	Consumption	Capital	Employment	Output
Standard Deviations				
Sample Economy	0.0146	0.0203	0.0100	0.0258
Model I Economy	0.0292	0.0241	0.0107	0.0397
	(0.0106)	(0.0066)	(0.0023)	(0.0112)
Model II Economy	0.1276	0.0425	0.0865	0.4648
	(0.1533)	(0.0238)	(0.1519)	(0.9002)
Correlation Coefficients				
Sample Economy				
Consumption	1.0000			
Capital Stock	0.4360	1.0000		
Employment	0.0039	−0.3002	1.0000	
Output	0.9692	0.5423	0.0202	1.0000
Model I Economy				
Consumption	1.0000			
	(0.0000)			
Capital Stock	0.7208	1.0000		
	(0.0920)	(0.0000)		
Employment	0.5138	−0.1842	1.0000	
	(0.1640)	(0.1309)	(0.0000)	
Output	0.9473	0.4855	0.7496	1.0000
	(0.0200)	(0.1099)	(0.1028)	(0.0000)
Model II Economy				
Consumption	1.0000			
	(0.0000)			
Capital Stock	0.6907	1.0000		
	(0.1461)	(0.0000)		
Employment	0.7147	0.3486	1.0000	
	(0.2319)	(0.4561)	(0.0000)	
Output	0.8935	0.5420	0.9130	1.0000
	(0.1047)	(0.2362)	(0.1312)	(0.00000)

Note: Numbers in parentheses are the corresponding standard errors.

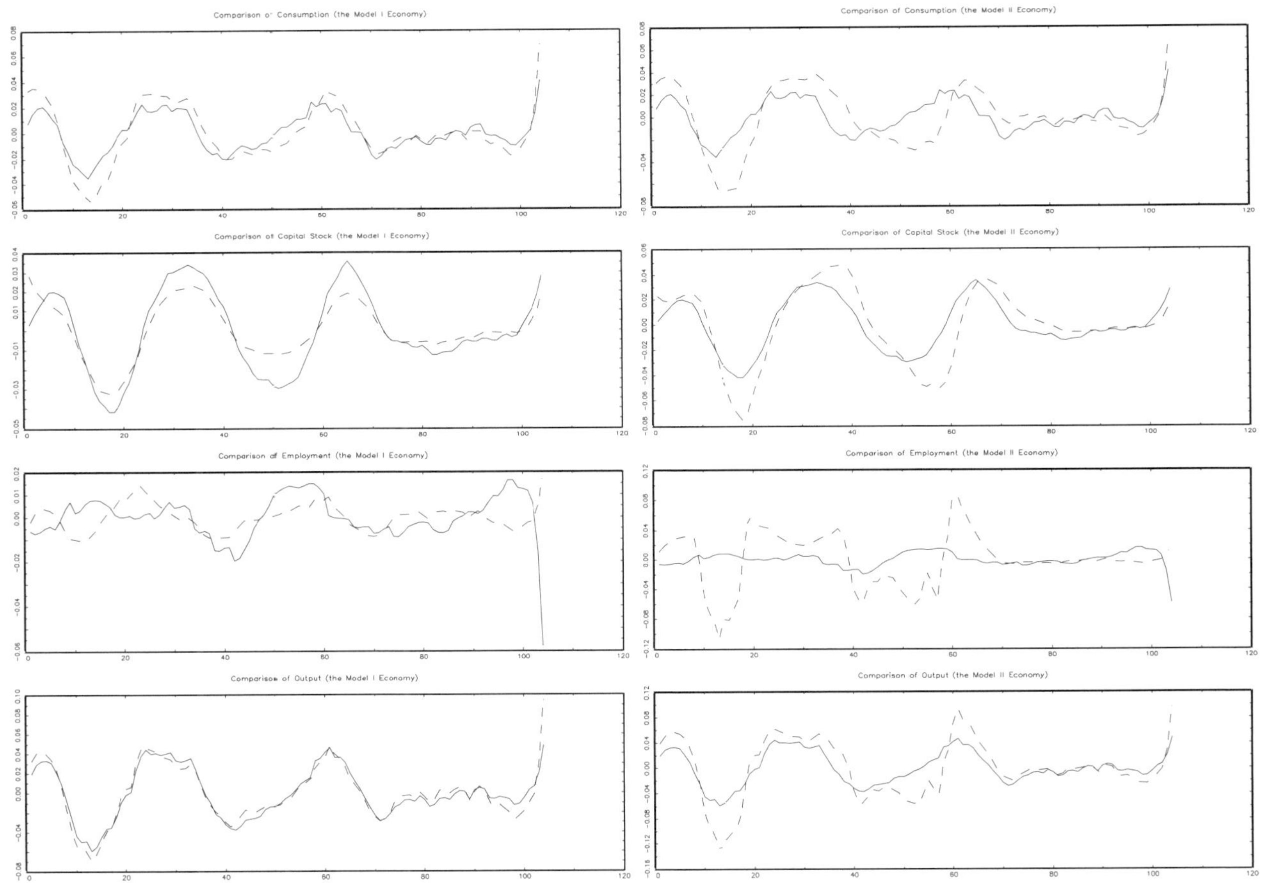

Figure 8.4. Simulated Economy and Sample Economy: Germany. Note: Solid line is sample economy; dashed line is simulated economy.

In contrast to the U.S. economy, we find some major differences. First, there is a difference in the variation of employment. The standard problem of excessive smoothness with respect to employment in the benchmark model does not hold for the German economy. This is likely due to the fact that employment is smooth in the German economy (see table 8.3 and figure 8.3). Also, the simulated labor supply in Germany is smoother than in the United States (see figure 8.5). In most labor market studies, the German labor market is often considered less flexible than the U.S. labor market. In particular, there are stronger influences of labor unions and legal restrictions on firms' hiring and firing decisions.[36] Such influences and legal restrictions give rise to a smoother employment series than in the United States. Such influences and legal restrictions—what Solow (1979) has termed the moral factor in the labor market—may also be viewed as a readiness to compromise, as our model III suggests. Those factors will give rise to a smooth employment series.

Further, if we look at the labor demand and supply in figure 8.5, the supply of labor is mostly the short side in the Germany economy, whereas in the U.S. economy demand is dominant in most periods. Here we must distinguish the actual labor supply and the "supply" that is specified by the decision rule in the standard model. It might reasonably be argued that because the intertemporal optimization is subject to budget constraints, the supply specified by the decision rule may only approximate the decisions of those households for which unemployment is not expected to pose a problem for their budgets. Such households are more likely to be currently employed and protected by labor unions and legal restrictions. In other words, currently employed labor, not those who are currently unemployed, decides about labor supply through the optimal decision rule. Such a shortcoming of a single representative intertemporal decision model could presumably be overcome by an intertemporal model with heterogeneous households.[37]

The second difference concerns the trend in employment growth and unemployment in the United States and Germany. So far, we have shown only that our model of a nonclearing labor market seems to match the variation in employment better than the standard RBC model does. This seems to be true particularly for the U.S. economy. We did not attempt to explain the trend of the unemployment rate for either the United States or Germany. The time series data (U.S., 1955.1–1983.1; Germany, 1960.1–1992.1) are from a period where the United States had higher—but falling—unemployment rates, whereas Germany had lower but rising unemployment rates. Since the end of the 1980s the unemployment rate in Germany has moved up considerably, partly due to the unification of Germany after 1989.

8.5 Differences in Labor Market Institutions

In section 8.2 we introduced rules that might be operative when there is a nonclearing labor market. In this respect, as our calibration in section

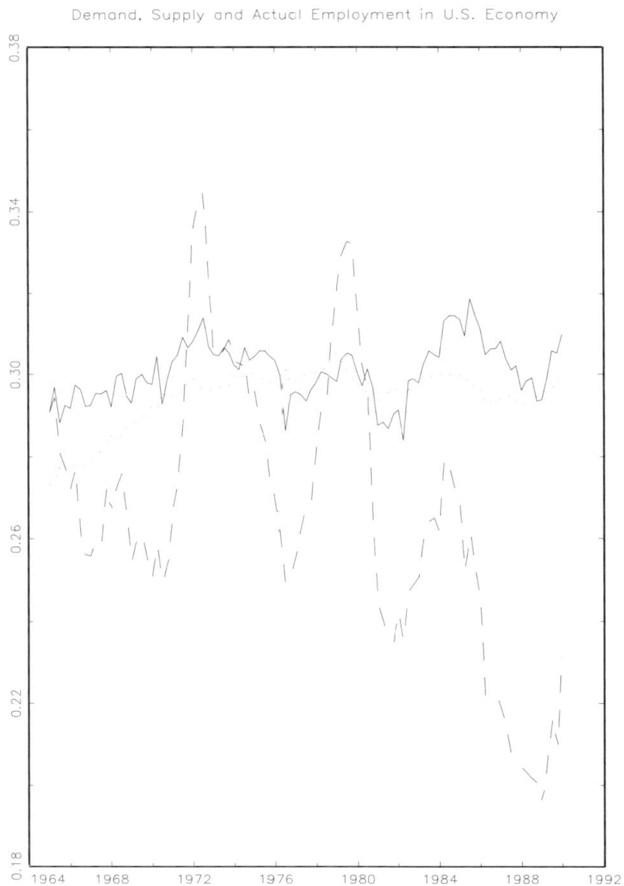

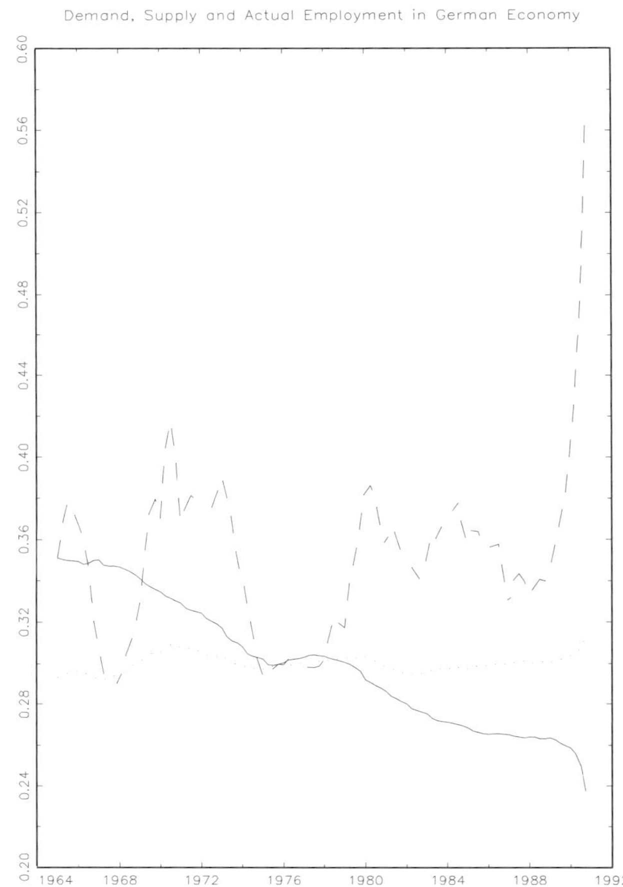

Figure 8.5. Comparison of Demand and Supply in the Labor Market. Note: Solid line is actual employment; dashed line is demand; dotted line is supply.

8.3 has shown, the most promising route to model and to match stylized facts of the labor market, through a microbased labor market behavior, is the compromising model. One may thereby pay attention to some institutional characteristics of the labor market presumed in our model.

The first is the way the agency representing the household sets the wage rate. If the household sets the wage rate as if it were a monopolistic competitor, then at this wage rate the household's willingness to supply labor is likely to be less than the market demand for labor unless the household sufficiently underestimates the market demand when it optimizes its wage setting. Such a way of wage setting may imply unemployment, and it is likely to be the institutional structure that gives the representative household (or the representative of the household, such as unions) the power to bargain with the firm in setting wages.[38] Yet, there could be other reasons why wages do not move to a labor market clearing level—such as an efficiency wage, an insider—outsider relationship, or wages determined by standards of fairness—as Solow (1979) has noted.

Second, the presence of union power is reflected not only in the wage bargaining process but also in the magnitude of ω. In addition, there can be labor market institutions, such as corporatist structures, also measured by our ω, that affect actual employment. Our ω expresses how much weight is given to the desired labor supply or desired labor demand. A small ω means that the agency representing the household has a high weight in determining the outcome of the employment compromise. A high ω means that the firm's side is stronger in employment negotiations. As our empirical estimations in Ernst et al. (2004) have shown, a low ω is very characteristic of Germany, France, and Italy, whereas a larger ω is found for the United States and the United Kingdom.[39]

Given the rather corporatist relationship of labor and the firm in some European countries, with some considerable labor market regulations through legislation and union bargaining (employment protection, unionization rate and bargaining coordination, extension of employment even if there is a shortfall of sales, etc.),[40] our ω may thus measure differences in labor market institutions between the United States and European countries. This was stated in the 1960s by Meyers: "One of the differences between the United States and Europe lies in our attitude toward layoffs. . . . When business falls off, he [the typical American employer] soon begins to think of reduction in work force. . . . In many other industrial countries, specific laws, collective agreements, or vigorous public opinion protects the workers against layoffs except under the most critical circumstances. Despite falling demand, the employer counts on retaining his permanent employees. He is obliged to find work for them to do. . . . These arrangements are certainly effective in holding down unemployment" (Meyers 1964, 180–181).

Thus, the major international difference causing employment variation arises less from real wage stickiness (due to the presence of unions and the extent and duration of contractual agreements between labor and the firm)[41] than from the degree to which compromising rules exist and which side dominates the compromising rule. A lower ω—defining, for example, the compromising rule in European countries—can show up as a difference in the variation of macroeconomic variables. This is demonstrated in section 8.4 for the German economy.

In section 8.4, employment and the efficiency wage (defined as real wage divided by production) are among the variables with the highest volatility in the U.S. economy. However, in the German economy they are the smoothest variables. Second, in the U.S. economy, the capital stock and temporary shock to technology are both relatively smooth. In contrast, they are more volatile in Germany. These results are likely to be due to the difference in employment volatility. The volatility of output must be absorbed by factors in the production function. If employment is smooth, the other two factors must be volatile.

Indeed, recent empirical studies do not seem to reveal much difference in real wage stickiness between Germany and the United States, although the German labor market is often considered less flexible.[42] Yet, there are differences in another sense. In Germany, there are stronger influences of labor unions and various legal restrictions on firms' hiring and firing decisions, and a shorter workweek (for the same pay).[43] Such influences and legal restrictions give rise to the smoother employment series in contrast to the United States. They may also be viewed as a readiness to compromise, as our model III suggests. Those factors will give rise to a lower ω and a smoother employment series.[44]

So far we have shown only that our model of a nonclearing labor market seems to match the variation in employment better than the standard RBC model does. However, we did not attempt to explain the secular trend of the unemployment rate for either the United States or Germany. We want to show how our model can be used to study the trend shift in employment. The time series data for table 8.3 (U.S., 1955.1–1983.1; Germany, 1960.1–1992.1) are from a period when the United States had higher—but falling—unemployment rates, whereas Germany had lower but rising unemployment rates. Since the end of the 1980s, the unemployment rate in Germany has considerably moved up, partly due to the unification of Germany after 1989.

One recent attempt to better fit the RBC model's predictions with labor market data has employed search and matching theory.[45] Informational or institutional search frictions may then explain the equilibrium unemployment rate and its rise. Yet, those models usually observe that there has been a shift in matching functions due to evolution of unemployment rates such as those experienced in Europe since the 1980s, and that the model itself fails to explain such a shift.[46]

In contrast to the literature on institutional frictions in the search and matching process, we think that the essential impact on the trend in the rate of unemployment seems to stem from both changes of household perferences and a changing trend in the technology shock.[47] Concerning the latter, as shown in chapters 5 and 9, the Solow residual, used in RBC models as the technology shock, greatly depends on endogenous variables (such as capacity utilization). Thus exogenous technology shocks constitute only a small fraction of the Solow residual. We thus might conclude that cyclical fluctuations in output and employment are not likely to be sufficiently explained by productivity shocks alone. Gali (1999) and Francis and Ramey (2001, 2003) have argued that other shocks, such as demand shocks, are important as well.

Yet, in the long run, the change in the trend of the unemployment rate is likely to be related to the long-run trend in the true technology shock. Empirical evidence on the role of lagging implementation and diffusion of new technology for low employment growth in Germany can be found in Heckman (2003) and Greiner et al. (2005). In the context of our model, this would have the effect that labor demand, given by equation (8.27), may fall short of labor supply, given by equation (8.24). This is likely to occur in the long run if the productivity Z_t in equation (8.27) starts tending to grow at a lower rate, which many researchers have maintained has happened in Germany, and other European countries, since the 1980s.[48] Yet, as recent research has stressed (e.g., Phelps 1997; Phelps and Zoega 1998), there have also been secular changes on the supply side of labor due to changes in preferences of households.[49] Some of those factors affecting the household's supply of labor have been discussed above.

8.6 Conclusions

Market clearing is a prominent feature in the standard RBC model that commonly presumes wage and price flexibility. In this chapter, we have introduced an adaptive optimization behavior and a multiple-stage decision process that, given wage stickiness, results in a nonclearing labor market in an otherwise standard stochastic dynamic model. A nonclearing labor market is, then, a result of different employment rules derived on the basis of a multiple-stage decision process. Calibrations have shown that such model variants will produce a higher volatility in employment, and thus fit the data significantly better than the standard model.[50]

Concerning international aspects of our study, we presume that different labor market institutions result in different weights defining the compromising rule. The results for European economies (e.g., Germany) in contrast to the United States are consistent with what has been found in many other empirical studies with regard to labor market institutions.

Finally, with respect to the trend of lower employment growth in some European countries compared with the United States since the 1980s, our model suggests that one has to study more carefully the secular forces affecting the supply and the demand of labor as modeled in our multiple-stage decision process (section 8.2). On the demand side for labor, the delay in catching up with information technology and a capital-biased technical change seem to have been major factors for the low employment growth in Germany and other countries in Europe.[51] On the other hand, there have also been changes in the preferences of households. Our study has provided a framework that allows following up on such issues. (For further discussion, see chapter 9.)

8.7 Appendix I: Wage Setting

Suppose that at the beginning of t, the household (with certain probability denoted as $1-\xi$) decides to set up a new wage rate w_1^*, given the data (A_t, k_t) and the sequence of expectations on $\{A_{t+i}\}_{i=1}^{\infty}$, where A_t and k_t are referred to as the technology and capital stock, respectively. If the household knows the production $f(A_t, k_t, n_t)$, where n_t is the labor effort, so that it also knows the firm's demand for labor, the decision problem of the household with regard to wage setting may be expressed as follows:

$$\max_{w_t^*, \{c_{t+i}\}_{i=0}^{\infty}} E_t \left[\sum_{i=0}^{\infty} (\xi\beta)^i \, U\left(c_{t+i}, n\left(w_t^*, k_{t+i}, A_{t+i}\right)\right) \right], \tag{8.28}$$

subject to

$$k_{t+i+1} = (1-\delta)k_{t+i} + f\left(A_{t+i}, k_{t+i}, n\left(w_t^*, k_{t+i}, A_{t+i}\right)\right) - c_{t+i}. \tag{8.29}$$

Above, ξ^i is the probability that the new wage rate w_t^* will still be effective in period $t+i$. Obviously, this probability will be reduced when i becomes larger. $U(\cdot)$ is the household's utility function, which depends on consumption c_{t+i} and the labor effort $n(w_t^*, k_{t+i}, A_{t+i})$. Here $n(w_t^*, k_{t+i}, A_{t+i})$ is the function of the firm's demand for labor, which is derived from the condition of marginal product equal to the wage rate:

$$w_t^* = f_n(A_{t+i}, k_{t+i}, n_{t+i}).$$

Although the decision is mainly about the choice of w_t^*, the sequence $\{c_{t+i}\}_{i=0}^{\infty}$ should also be considered for dynamic optimization. Of course there is no guarantee that the household will actually implement $\{c_{t+i}\}_{i=0}^{\infty}$. However, as argued by recent New Keynesian literature, there is only a certain probability (due to the adjustment cost of changing the wage) that the household will set a new wage rate in period t. Therefore, the observed wage dynamics w_t may follow Calvo's updating scheme:

$$w_t = (1-\xi)w_t^* + \xi w_{t-1}.$$

Such a wage indicates that there exists a gap between the optimum wage w_t^* and the observed wage w_t (see also Gong and Semmler, 2005).

In recent New Keynesian literature, where the wage is set in a way similar to what we have discussed here, the concept of a nonclearing labor market has somehow disappeared. The household is assumed to supply the labor effort according to the market demand at the existing wage rate, and therefore does not seem to face the problem of excess demand or supply. Instead, what New Keynesian economists are concerned with is the gap between the optimum price and the actual price, the existence of which is caused by the adjustment cost of changing prices. Corresponding to the gap between optimum and actual price, there is also a gap between optimum output and actual output.

Some clarifications may be obtained by referring to a static version of our view on the working of the labor market. In figure 8.6, the supplier (or the household, in the labor market case) first (say, at the beginning of period 0) sets its price optimally according to the expected demand curve D_0. Let us denote this price as w_0. Consider now the situation that the supplier's expectation of demand is not fulfilled. Instead of n_0, the market demand at w_0 is n'. In this case, the household may reasonably believe that the demand curve should be D', and therefore the optimum price should be w^* and the optimum supply should be n^*. Yet, due to the adjustment cost of changing prices, the supplier may stick to w_0. This produces the gaps between

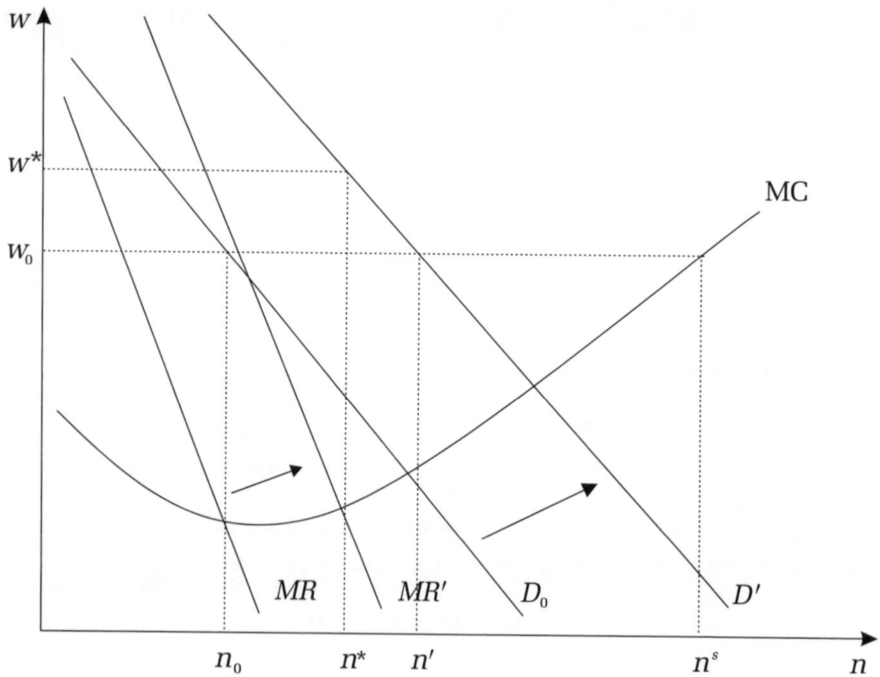

Figure 8.6. A Static Version of the Working of the Labor Market

optimum price w^* and actual price w_0, and between optimum supply n^* and actual supply n'.

However, the existence of price and output gaps does not exclude the existence of a disequilibrium or nonclearing market. New Keynesian literature presumes that at the existing wage rate, the household supplies labor effort regardless of what the market demand for labor is. Note that in figure 8.6 the household's willingness to supply labor is n^s. In this context the marginal cost curve, MC, can be interpreted as marginal disutility of labor, which has an upward slope because we use the standard log-utility function, as in the RBC literature. This means that the household's supply of labor will be restricted by a wage rate below, or equal to, the marginal disutility of work. If we define the labor market demand and supply in a standard way (i.e., at the given wage rate), there is a firm's willingness to demand labor and the household's willingness to supply labor, and a nonclearing labor market can be a very general phenomenon. This indicates that even if there are no adjustment costs, so that the household can adjust the wage rate in every t (so that there are no price and quantity gaps), the disequilibrium in the labor market may still exist.

8.8 Appendix II: Adaptive Optimization and Consumption Decision

For the problems (8.14)–(8.16), we define the Lagrangian:

$$L = E_t\left\{\left[\log c_t^d + \theta \log(1 - n_t)\right] + \right.$$

$$\left.\lambda_t\left[k_{t+1}^s - \frac{1}{1+\gamma}\left[(1-\delta)k_t^s + f\left(k_t^s, n_t, A_t\right) - c_t^d\right]\right]\right\}$$

$$+ E_t\left\{\sum_{i=1}^{\infty}\beta^i\left[\log\left(c_{t+i}^d\right) + \theta\log\left(1 - n_{t+i}^s\right)\right] + \right.$$

$$\left.\beta^i\lambda_{t+i}\left[k_{t+1+i}^s - \frac{1}{1+\gamma}\left[(1-\delta)k_{t+i}^s + f\left(k_{t+i}^s, n_{t+i}^s, A_{t+i}\right) - c_{t+i}^d\right]\right]\right\}.$$

Since the decision is only about c_t^d, we take the partial derivatives of L with respect to c_t^d, k_{t+1}^s, and λ_t. This gives us the following first-order condition:

$$\frac{1}{c_t^d} - \frac{\lambda_t}{1+\gamma} = 0, \tag{8.30}$$

$$\frac{\beta}{1+\gamma}E_t\left\{\lambda_{t+1}\left[(1-\delta) + (1-\alpha)A_{t+1}\left(k_{t+1}^s\right)^{-\alpha}\left(n_{t+1}^s\bar{N}/0.3\right)^{\alpha}\right]\right\} = \lambda_t, \tag{8.31}$$

$$k_{t+1}^s = \frac{1}{1+\gamma}\left[(1-\delta)k_t^s + A_t\left(k_t^s\right)^{1-\alpha}\left(n_t\bar{N}/0.3\right)^{\alpha} - c_t^d\right]. \tag{8.32}$$

In deriving the decision rules as expressed in (8.23) and (8.24), we postulated

$$\lambda_{t+1} = Hk_{t+1}^s + QA_{t+1} + h, \tag{8.33}$$

$$n_{t+1}^s = G_{21}k_{t+1}^s + G_{22}A_{t+1} + g_2, \tag{8.34}$$

where H, Q, h, G_{21}, G_{22}, and g_2 have all been resolved previously in the household optimization program. We therefore obtain, from (8.33) and (8.34),

$$E_t\lambda_{t+1} = Hk_{t+1}^s + Q(a_0 + a_1A_t) + h, \tag{8.35}$$

$$E_tn_{t+1}^s = G_2k_{t+1}^s + D_2(a_0 + a_1A_t) + g_2. \tag{8.36}$$

Our next step is to linearize (8.30)–(8.32) around the steady states. Suppose that they can be written as

$$F_{c1}c_t + F_{c2}\lambda_t + f_c = 0, \tag{8.37}$$

$$F_{k1}E_t\lambda_{t+1} + F_{k2}E_tA_{t+1} + F_{k3}k_{t+1}^s + F_{k4}E_tn_{t+1}^s + f_k = \lambda_t, \tag{8.38}$$

$$k_{t+1}^s = Ak_t + WA_t + C_1c_t^d + C_2n_t + b. \tag{8.39}$$

Expressing $E_t\lambda_{t+1}$, $E_tn_{t+1}^s$, and E_tA_{t+1} in terms of (8.35), (8.36), and $a_0+a_1A_t$, respectively, from (8.38) we obtain

$$\kappa_1k_{t+1}^s + \kappa_2A_t + \kappa_0 = \lambda_t, \tag{8.40}$$

where, in particular,

$$\kappa_0 = F_{k1}(Qa_0 + h) + F_{k2}a_0 + F_{k4}(G_{22}a_0 + g_2) + f_k, \tag{8.41}$$

$$\kappa_1 = F_{k1}H + F_{k3} + F_{k4}G_{21}, \tag{8.42}$$

$$\kappa_2 = F_{k1}Qa_1 + F_{k2}a_1 + F_{k4}G_{22}a_1. \tag{8.43}$$

Using (8.37) to express λ_t in (8.40), we further obtain

$$\kappa_1k_{t+1}^s + \kappa_2A_t + \kappa_0 = -\frac{F_{c1}}{F_{c2}}c_t^d - \frac{f_c}{F_{c2}}, \tag{8.44}$$

which is equivalent to

$$k_{t+1}^s = -\frac{\kappa_2}{\kappa_1}A_t - \frac{F_{c1}}{F_{c2}\kappa_1}c_t^d - \frac{\kappa_0}{\kappa_1} - \frac{f_c}{F_{c2}\kappa_1}. \tag{8.45}$$

Comparing the right sides of (8.39) and (8.45) will allow us to solve c_t^d as

$$c_t^d = -\left(\frac{F_{c1}}{F_{c2}\kappa_1} + C_1\right)^{-1}\left[Ak_t + \left(\frac{\kappa_2}{\kappa_1} + W\right)A_t + C_2 n_t + \left(b + \frac{\kappa_0}{\kappa_1} + \frac{f_c}{F_{c2}\kappa_1}\right)\right].$$

From the latter equation we can infer that actual employment, n_t, also determines consumption, in addition to the technology shock, A_t, and the capital stock, k_t, which are obtained from the optimal solution.

8.9 Appendix III: Welfare Comparison of the Model Variants

In this appendix we undertake a welfare comparison of our model variants. Here we concentrate solely on the model for the United States. We could also undertake a welfare comparison for the United States and Europe (for either of the model variants). Although the per capita output in the United States is higher than in Europe, the hours worked in Europe are lower. Since the Europeans enjoy more leisure time, the welfare turns out to be roughly the same, a result different from the one obtained by Prescott (2003), who attributes the lower per capita income in Europe to marginal tax rates.

Concentrating on the welfare comparison of different model variants for one country, we follow Ljungqvist and Sargent (1998) and compute the welfare implications of the different model variants.[52] However, whereas they concentrate on the steady state, we also compute the welfare outside the steady state. We restrict our welfare analysis to the U. S. model variants. It is sufficient to consider only the equilibrium (benchmark) model and the two models with a nonclearing labor market. They are given by simulated economies I, II, and III. A likely conjecture is that the benchmark model should always be superior to the two variants because the decisions on labor supply—which are optimal for the representative agent—are realized in all periods.

However, we believe that this may not generically be the case. The point here is that the model specification in variants II and III is somewhat different from the the benchmark model because of the distinction between expected and actual moments with respect to our state variable, the capital stock. In the models of a nonclearing market, the representative agent may not rationally expect those moments of the capital stock. The expected moments are represented by equation (8.5), while the actual moments are expressed by equation (8.19). They are not necessarily equal unless the labor efforts of those two equations are equal. Also, in addition to A_t, there is another external variable w_t, entering into the models, which will affect the labor employed (via demand for labor) and, hence, the welfare performance. The welfare result due to these changes in the specification may therefore deviate from what one would expect.

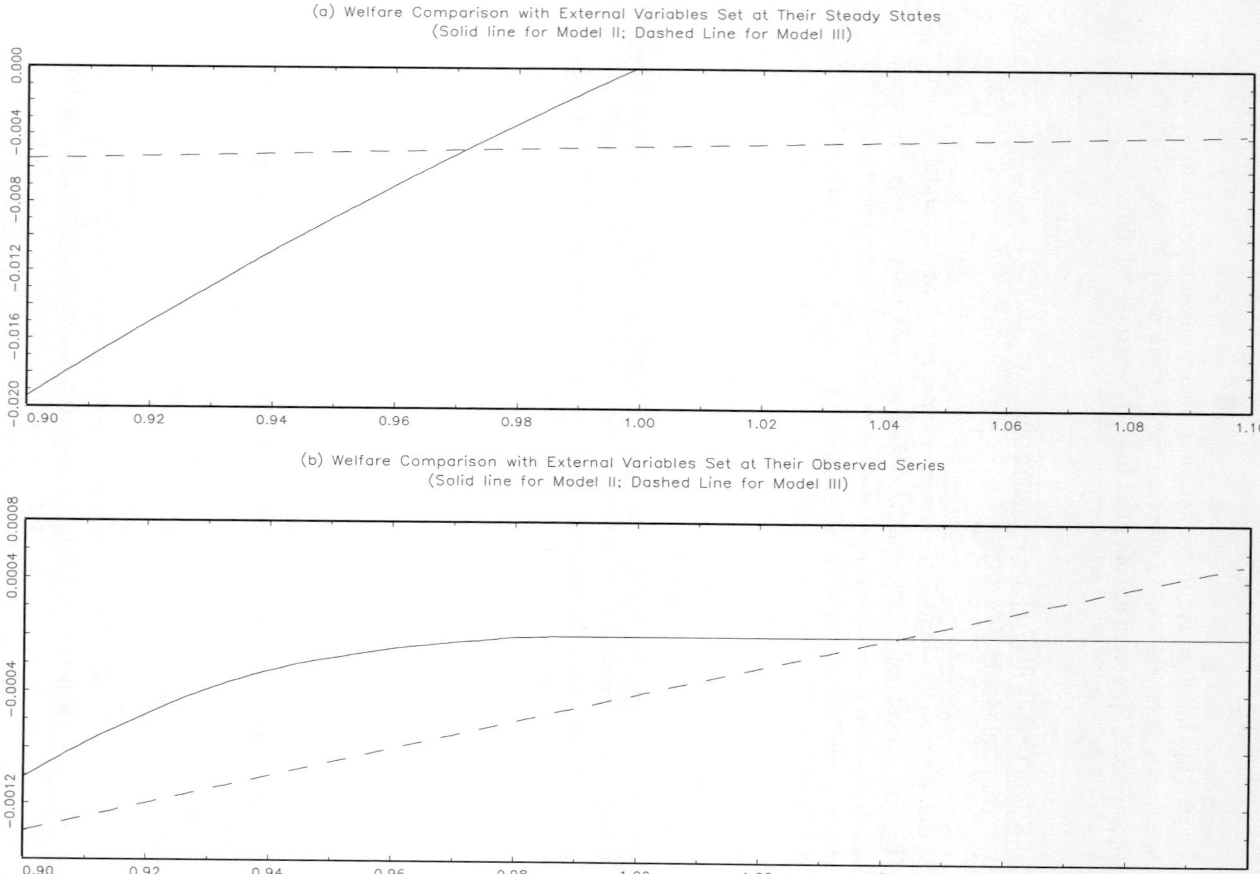

Figure 8.7. **Welfare Comparison of Models II and III. Note: In each graph, solid line is model II; dashed line is model III.**

Our exercise here is to compute the values of the objective function for all three models, given the sequence of our two decision variables, consumption and employment. For our model variants with nonclearing labor market, we use realized employment, rather than the decisions on labor supply, to compute the utility functional. More specifically, we calculate V, where

$$V \equiv \sum_{t=0}^{\infty} \beta^t U(c_t, n_t),$$

where $U(c_t, n_t)$ is given by $\log(c_t)+\theta \log(1-n_t)$. This exercise is conducted for different initial conditions of k_t denoted by k_0. We choose the different k_0's based on the grid search around the steady state of k_t. Obviously, the value of V for any given k_0 will also depend on the external variables A_t and w_t (though in the benchmark model, only A_t appears). We consider two different ways to treat these external variables. One is to set both external variables at their steady state levels for all t. The other is to employ their observed series entering into the computation. Figure 8.7 provides the welfare comparison of the two versions.

In figure 8.7, the percentage deviations of V from the corresponding values of the benchmark model are plotted for both model II and model III given for various k_0's around the steady states. The various k_0's are expressed in terms of the percentage deviation from the steady state k_t.

It is not surprising that in most cases the benchmark model is the best in its welfare performance, since most of the values are negative. However, it is important to note that the deviations from the benchmark model are very small. Similar results have been obtained by Ljungqvist and Sargent (1998); however, they compare only the steady states. However, the benchmark model is not always the best one. When k_0 is sufficiently high, close to or higher than the steady state of k_t, the deviations become 0 for model II. Furthermore, when using observed external variables, model III will be superior in its welfare performance when k_0 is larger than its steady state. (See lower part of the figure 8.7.)

9

Monopolistic Competition, Nonclearing Markets, and Technology Shocks

9.1 Introduction

In chapter 8 we found that if we introduce some non–Walrasian features into an intertemporal decision model with the household's wage setting, sluggish wage and price adjustments, and adaptive optimization, the labor market may not be cleared. This model naturally generates higher volatility of employment and a low correlation between employment and consumption. Those results appear to be strongly supported by empirical evidence. We relate our approach of a nonclearing labor market to the theory of monopolistic competition in the product market as developed in New Keynesian economics.

In many respects, the specifications in this chapter are the same as for the model of chapter 8. We still follow the assumptions with respect to ownership, adaptive optimization, and nonclearing labor markets. The assumption of reopening of the market is also adopted. This is necessary for a model with nonclearing markets where adjustments should take place in real time.

9.2 The Model

As mentioned in chapter 8, price and wage stickiness is an important feature in New Keynesian literature. Keynes (1936) attributed strong stabilizing effects to wage stickiness. Recent literature uses monopolistic competition theory to give a foundation to nominal stickiness.

Since both households and firms make their quantity decisions on the basis of given prices, including the output price p_t, the wage rate w_t, and the rental rate of capital stock r_t, we will first discuss how the period t prices are determined at the beginning of period t in our model.

Here again, as in the model of chapter 8, there are three commodities. One of them should serve as a numeraire, which we assume to be the output. Therefore, the output price p_t always equals 1. This indicates that the wage w_t and the rental rate of capital stock r_t are measured in terms of the physical units of output.[1] The rental rate of capital r_t is assumed to be adjustable and to clear the capital market. We can then ignore its setting. Here, we follow all the specifications on setting prices and wages presented in Chapter 8, section 8.2.1.

9.2.1 The Household's Desired Transactions

When prices, including wages, have been set, the household will express its desired demand and supply. We define the household's willingness as the demand and supply that can allow the household to obtain the maximum utility on the condition that such demand and supply can be realized at the given set of prices. We can express this as a sequence of output demand and factor supplies $\left\{c_{t+i}^d, i_{t+i}^d, n_{t+i}^s, k_{t+i+1}^s\right\}_{i=0}^{\infty}$, where i_{t+i} refers to investment. Here we have used the superscripts d and s to refer to the agent's desired demand and supply. The decision problem for the household to derive its desired demand and supply is very similar to that in chapter 8 and can be formulated as

$$\max_{\left\{c_{t+i}^d, n_{t+i}^s\right\}_{i=0}^{\infty}} E_t\left[\sum_{i=0}^{\infty} \beta^i U\left(c_{t+i}^d, n_{t+i}^s\right)\right], \tag{9.1}$$

subject to

$$k_{t+i+1}^s = (1-\delta)k_{t+i}^s + f\left(k_{t+i}^s, n_{t+i}^s, A_{t+i}\right) - c_{t+i}^d. \tag{9.2}$$

The notation was defined in chapter 8. For the given technology sequence $\{A_{t+i}\}_{i=0}^{\infty}$, the solution of the optimization problem can be written as

$$c_{t+i}^d = G_c\left(k_{t+i}^s, A_{t+i}\right) \tag{9.3}$$

$$n_{t+i}^s = G_n\left(k_{t+i}^s, A_{t+i}\right). \tag{9.4}$$

Although the solution appears to be a sequence $\left\{c_{t+i}^d, n_{t+i}^s\right\}_{i=0}^{\infty}$, only (c_t^d, n_t^s), and (i_t^d, k_t^s), where $i_t^d = f(k_t^s, n_t^s, A_t) - c_t^d$ and $k_t^s = k_t$, are actually carried by the household into the market for exchange, due to our assumption of reopening the market.

9.2.2 The Quantity Decisions of the Firm

The problem of our representative firm in period t is to choose the current input demand and output supply (n_t^d, k_t^d, y_t^s) so as to maximize the current profit. However, in this chapter, we no longer assume that the product market is in perfect competition. Instead, we assume that our representative firm behaves as a monopolistic competitor, and therefore it should face a perceived demand curve for its product. (See the discussion in chapter 8.) Thus, given the output price, which will always be 1 (since it serves as a numeraire), the firm has a perceived constraint on the market demand for its product. We denote this perceived demand as $\widehat{y}_t$.

On the other hand, given the prices of output, labor, and capital stock $(1, w_t, r_t)$, the firm should also have its own desired supply y_t^*. This desired supply is the amount that allows the firm to obtain a maximum profit on the assumption that all its output can be sold. Obviously, if the expected demand $\widehat{y}_t$ is less than the firm's desired supply y_t^*, the firm will choose $\widehat{y}_t$. Otherwise, it will simply follow the short-side rule to choose y_t^*, as in the New Keynesian model.

Thus, for our representative firm, the optimization problem can be expressed as

$$\max \left[\min \left(\widehat{y}_t, y_t^* \right) - r_t k_t^d - w_t n_t^d \right], \tag{9.5}$$

subject to

$$\min \left(\widehat{y}_t, y_t^* \right) = f(A_t, k_t, n_t). \tag{9.6}$$

For the regular condition on the production function, the solutions should satisfy

$$k_t^d = f_k(r_t, w_t, A_t, \widehat{y}_t) \tag{9.7}$$

$$n_t^d = f_n(r_t, w_t, A_t, \widehat{y}_t), \tag{9.8}$$

where r_t and w_t are, respectively, the prices (in real terms) of capital and labor. (The details will be presented in the appendix to this chapter.)

We now consider the transactions in our three markets. Let us first consider the two factor markets.

9.2.3 Transactions in the Factor Market

Since the rental rate of capital stock r_t is adjusted to clear the capital market when the market is reopened in period t, we have

$$k_t = k_t^s = k_t^d. \tag{9.9}$$

Due to the monopolistic wage setting and the sluggish wage adjustment, there is no reason to believe that the labor market will be cleared. (See the discussion in chapter 8.) Therefore, we again define a realization rule with regard to actual employment. As was discussed in chapter 8, the rule that has been used most frequently is the short-side rule:

$$n_t = \min\left(n_t^d, n_t^s\right).$$

Thus, when a disequilibrium occurs, only the short side of demand and supply will be realized. Another important rule that was discussed in chapter 8 is the compromising rule: When disequilibrium occurs in the labor market, both firms and workers have to compromise. We again formulate this rule as

$$n_t = \omega n_t^d + (1-\omega)n_t^s, \tag{9.10}$$

where $\omega \in (0, 1)$. Our study in chapter 8 indicates that the short-side rule seems to be empirically less satisfying than the compromising rule. Therefore, in this chapter we consider only the compromising rule.

9.2.4 The Transactions in the Product Market

After the transactions in the factor market have been carried out, the firm will engage in production. The result is the output supply, which is given by

$$y_t^s = f(k_t, n_t, A_t). \tag{9.11}$$

Equation (9.11) indicates that the firm's actual output is not necessarily constrained by equation (9.6), and therefore one may argue that the output determination does eventually follow the Keynesian way—that is, output is constrained by demand. However, the Keynesian way of determining output is still reflected in the firm's demand for inputs, capital, and labor (see (9.7) and (9.8)). On the other hand, if the output is still constrained by (9.6), one may encouter the difficulty either in terms of feasibility when y_t^s in (9.11) is less than $\min(\widehat{y}_t, y_t^*)$ or in terms of inefficiency when y_t^s is larger than $\min(\widehat{y}_t, y_t^*)$.[2]

Given that the output is determined by (9.11), the transaction needs to be carried out with respect to y_t^s. It is important to note here that when disequilibrium occurs in the labor market, the consumption plan expressed by (9.3) becomes invalid due to the improper rule of capital accumulation (9.2) for deriving the plan. Therefore, the household will construct a new plan:

$$\max_{\left(c_t^d\right)} U\left(c_t^d, n_t\right) + E_t\left[\sum_{i=0}^{\infty} \beta^i U\left(c_{t+i}^d, n_{t+i}^s\right)\right] \tag{9.12}$$

$$\text{s.t. } k^s_{t+1} = \frac{1}{1+\gamma}\left[(1-\delta)k^s_t + f(k_t, n_t, A_t) - c^d_t\right] \tag{9.13}$$

$$k^s_{t+i+1} = \frac{1}{1+\gamma}\left[(1-\delta)k^s_{t+i} + f\left(k^s_{t+i}, n^s_{t+i}, A_{t+i}\right) - c^d_{t+i}\right], \tag{9.14}$$

$$i = 1, 2, \ldots$$

Above, $k_t = k^s_t$ is given by (9.9) and n_t is given by (9.10), with n^s_t and n^d_t implied by (9.4) and (9.8), respectively. As was demonstrated in chapter 8, the solution to this further step in the optimization problem can be written in terms of the following equation:

$$c^d_t = G_{c2}(k_t, A_t, n_t), \tag{9.15}$$

where it is indicated that actual consumption demand depends not only on the optimal decision with respect to the paths of A_t and k_t but also on actual employment, n_t. Given this consumption plan, the product market could be cleared if the household demands the amount $f(k_t, n_t, A_t) - c^d_t$ for investment. In this case, therefore, c^d_t in (9.15) should also be the realized consumption.

9.3 Estimation and Calibration for the U.S. Economy

9.3.1 The Empirically Testable Model

This section provides an empirical study of the theoretical model presented above, which, in order to make it empirically more realistic, has to include economic growth.

Let K_t denote capital stock; N_t, per capita working hours; Y_t, output; and C_t, consumption. Assume the capital stock in the economy follows the transition law:

$$K_{t+1} = (1-\delta)K_t + A_t K_t^{1-\alpha}(N_t X_t)^\alpha - C_t, \tag{9.16}$$

where δ is the depreciation rate; α is the share of labor in the production function $F(\cdot) = A_t K_t^{1-\alpha}(N_t X_t)^\alpha$; A_t is the temporary shock in technology; and X_t is the permanent shock that follows a growth rate γ. Dividing both sides of equation (9.16) by X_t, we obtain

$$k_{t+1} = \frac{1}{1+\gamma}\left[(1-\delta)k_t + A_t k_t^{1-\alpha}(n_t \bar{N}/0.3)^\alpha - c_t\right], \tag{9.17}$$

where $k_t \equiv K_t/X_t$, $c_t \equiv C_t/X_t$, and $n_t \equiv 0.3N_t/\bar{N}$, with $\bar{N}$ the sample mean of N_t. The above formulation also indicates that the function $f(\cdot)$ in section 9.2.4 may take the form

$$f(\cdot) = A_t k_t^{1-\alpha} (n_t \bar{N} / 0.3)^{\alpha}. \tag{9.18}$$

With regard to the household preference, we assume that the utility function takes the form

$$U(c_t, n_t) = \log c_t + \theta \log (1 - n_t). \tag{9.19}$$

The temporary shock A_t may follow an AR(1) process:

$$A_{t+1} = a_0 + a_1 A_t + \epsilon_{t+1}, \tag{9.20}$$

where ϵ_t is an independently and identically distributed (i.i.d.) innovation: $\epsilon_t \sim N(0, \sigma_\epsilon^2)$.

Finally, we assume that the output expectation $\widehat{y}_t$ is equal to y_{t-1}, that is,

$$\widehat{y}_t = y_{t-1}, \tag{9.21}$$

where $y_t = Y_t / X_t$, so that the expectation is fully adaptive to the actual output in the last period.[3]

9.3.2 The Data-Generating Process

For our empirical assessment, we consider two model variants: the standard model, as a benchmark for comparison, and our model, with monopolistic competition and nonclearing labor market. Specifically, we will call the benchmark model "model I" and the model with monopolistic competition "model IV" (to distinguish it from models II and III in chapter 8).

For the benchmark dynamic optimization model, model I, the data-generating process includes (9.17) and (9.20) as well as

$$c_t = G_{11} A_t + G_{12} k_t + g_1 \tag{9.22}$$

$$n_t = G_{21} A_t + G_{22} k_t + g_2. \tag{9.23}$$

Here, (9.22) and (9.23) are the linear approximations to (9.3) and (9.4). The coefficients G_{ij} and $g_i (i = 1, 2$ and $j = 1, 2)$ are the complicated functions of the model's structural parameters, α, β, and δ, among others. They are computed by the numerical algorithm, using the linear-quadratic approximation method.[4]

To define the data-generating process for our model with monopolistic competition and nonclearing labor market, model IV, we first modify (9.23) as

$$n_t^s = G_{21} A_t + G_{22} k_t + g_2. \tag{9.24}$$

However, the equilibrium in the product market indicates that c_t^d in (9.15) should be equal to c_t, and therefore this equation can, as shown in chapter 8, be approximated by

$$c_t = G_{31}A_t + G_{32}k_t + G_{33}n_t + g_3. \tag{9.25}$$

The computation of coefficients g_3 and G_{3j}, $j = 1, 2, 3$, are the same as in chapter 8.

Next we consider the demand for labor n_t^d derived from the firm's optimization problem (9.5)–(9.8), which will be augmented by the growth factor for our empirical test. The following proposition concerns the derivation of n_t^d.

Proposition 7 *When the capital market is cleared, the firm's demand for labor can be expressed as*

$$n_t^d = \begin{cases} \left(\frac{0.3}{\bar{N}}\right)\left(\frac{\hat{y}_t}{A_t}\right)^{1/\alpha}\left(\frac{1}{k_t}\right)^{(1-\alpha)/\alpha} & \text{if } \hat{y}_t < y_t^* \\ \frac{0.3}{\bar{N}}\left(\frac{\alpha A_t Z_t}{w_t}\right)^{\frac{1}{1-\alpha}} k_t & \text{if } \hat{y}_t \geq y_t^* \end{cases} \tag{9.26}$$

where

$$y_t^* = (\alpha A_t Z_t / w_t)^{\alpha/(1-\alpha)} k_t A_t. \tag{9.27}$$

The first n_t^d in (9.26) corresponds to the condition that the expected demand is less than the firm's desired supply, and the second n_t^d, to the condition otherwise. The proof of this proposition is provided in the appendix to this chapter. Thus, for model IV, the data-generating process includes (9.17), (9.20), (9.10), (9.24), (9.25), (9.26), and (9.21), with w_t given by the observed wage rate. Again, we do not need to attempt to give the actually observed sequence of wages a further theoretical foundation. For our purpose it suffices to take the empirically observed series of wages.

9.3.3 The Data and the Parameters

Here we employ only time series data of the U.S. economy. To calibrate the models, we first specify the structural parameters. There are ten structural parameters in model IV: $a_0, a_1, \sigma_\varepsilon, \gamma, \mu, \alpha, \beta, \delta, \theta$, and ω. All these parameters, except for the ω, are essentially the same as those employed in chapter 8 (see table 8.1). Here we choose ω to be 0.5203. This is estimated according to our new model by minimizing the residual sum of squares between actual employment and the model-generated employment. The estimation is again executed by a conventional algorithm, the grid search. Again we need a rescaling of the wage series in the estimation of ω.[5]

Table 9.1. Calibration of the Model Variants

	Consumption	Capital	Employment	Output
Standard Deviations				
Sample Economy	0.0081	0.0035	0.0165	0.0156
Model I Economy	0.0091	0.0036	0.0051	0.0158
	(0.0012)	(0.0007)	(0.0006)	(0.0021)
Model IV Economy	0.0071	0.0058	0.0237	0.0230
	(0.0015)	(0.0018)	(0.0084)	(0.0060)
Correlation Coefficients				
Sample Economy				
Consumption	1.0000			
Capital Stock	0.1741	1.0000		
Employment	0.4604	0.2861	1.0000	
Output	0.7550	0.0954	0.7263	1.0000
Model I Economy				
Consumption	1.0000			
	(0.0000)			
Capital Stock	0.2043	1.0000		
	(0.1190)	(0.0000)		
Employment	0.9288	−0.1593	1.0000	
	(0.0203)	(0.0906)	(0.0000)	
Output	0.9866	0.0566	0.9754	1.0000
	(0.0033)	(0.1044)	(0.0076)	(0.0000)
Model IV Economy				
Consumption	1.0000			
	(0.0000)			
Capital Stock	0.3878	1.0000		
	(0.1515)	(0.0000)		
Employment	0.4659	0.0278	1.0000	
	(0.1424)	(0.1332)	(0.0000)	
Output	0.8374	0.0369	0.8164	1.0000
	(0.0591)	(0.0888)	(0.1230)	(0.0000)

Note: Numbers in parentheses are the corresponding standard deviations.

9.3.4 Calibration

Table 9.1 provides the results of our calibrations from 5,000 stochastic simulations. This result is further confirmed by figure 9.1, where a one-time simulation with the observed innovation A_t is presented.[6] All time series are detrended by the HP-filter.

9.3.5 The Labor Market Puzzle

Despite the bias toward model I economy, due to the selection of the structural parameters, the labor effort is much more volatile than in the model

I economy, the benchmark model. Indeed, in comparison with the benchmark model, the volatility of labor effort in our model IV economy is much increased; if anything, it is too high. This result is, however, not surprising because the agents face two constraints—one in the labor market and one in the product market. Also, the excessive correlation between labor and consumption has been weakened.

Further evidence on the better fit of our model IV economy—as concerns the volatility of the macroeconomic variables—is demonstrated in figure 9.1, where the graphs show, from top to bottom, actual (solid line) and simulated data (dotted line) for consumption, capital stock, employment, and output. The two columns, from left to right, represent model I and model IV economies, respectively. As can be seen, the employment series in the model IV economy fit the data better than the model I economy.

This resolution to the labor market puzzle should not be surprising because we specify the structure of the labor market in essentially the same way as in chapter 8. However, in addition to the labor market disequilibrium specified in chapter 8, in this chapter we allow for monopolistic competition in the product market. In addition to impacting the volatility of labor effort, this may provide the possibility to resolve the technology puzzle arising in the market-clearing RBC model.

9.3.6 The Technology Puzzle

Economic literature often discusses technology in terms of its persistent and temporary effects on the economy. One possible way to investigate the persistent effect in our models is to look at the steady states. Given that at the steady state all the markets will be cleared, our model IV economy should have the same steady state as the benchmark model. For the convenience of our discussion, we rewrite these steady states in the following equations (see the proof of proposition 4 in chapter 4):

$$\overline{n} = \alpha\phi / \left[(\alpha + \theta)\phi - (\delta + \gamma)\theta\right]$$

$$\overline{k} = \overline{A}^{1/\alpha}\phi^{-1/\alpha}\overline{n}\left(\overline{N}/0.3\right)$$

$$\overline{c} = (\phi - \delta - \gamma)\overline{k}$$

$$\overline{y} = \phi\overline{k},$$

where

$$\phi = \left[(1+\gamma) - \beta(1-\delta)\right] / \beta(1-\alpha).$$

The above equation reveals that technology has a positive, persistent effect on output, consumption, and capital stock,[7] yet 0 effect on employment.

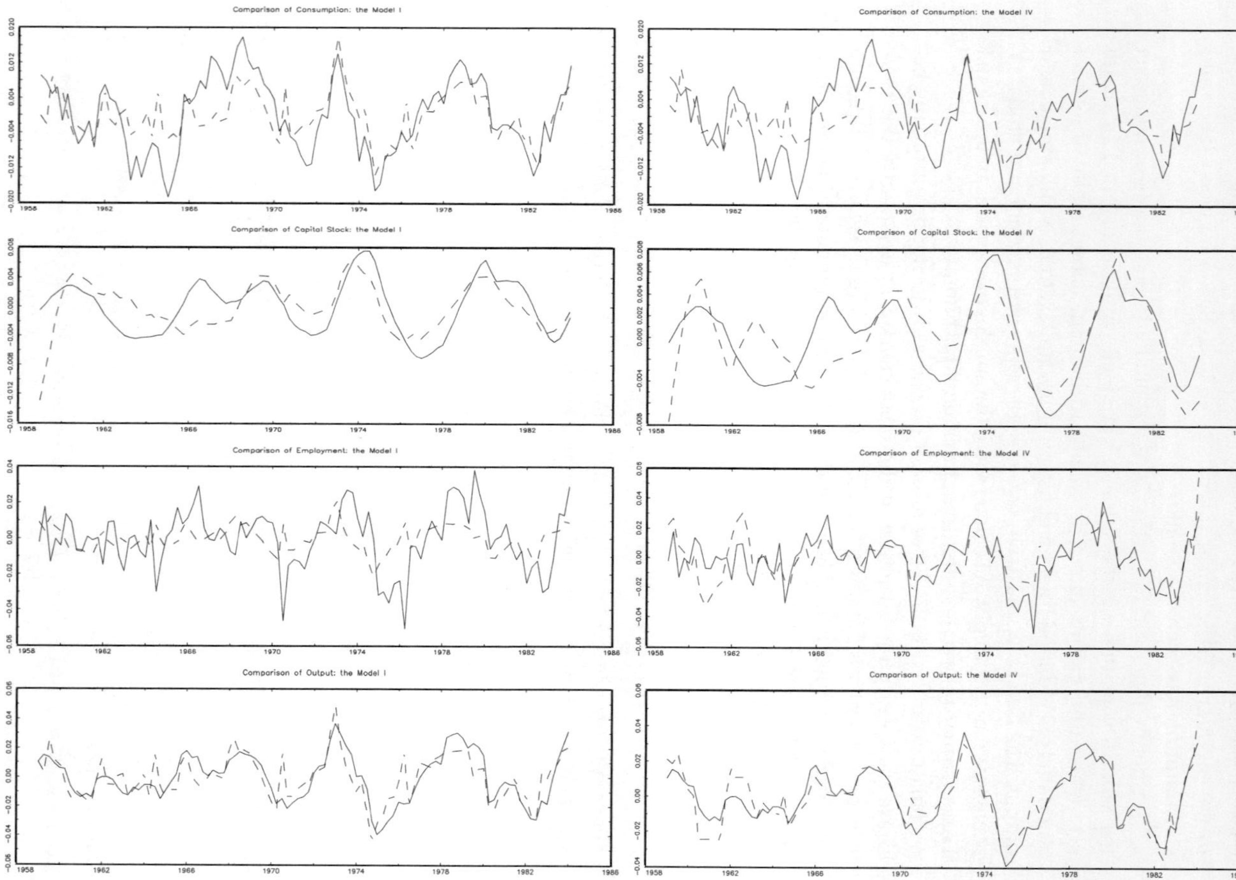

Figure 9.1. Simulated Economy versus U.S. Economy. Note: Solid line is sample economy; dashed line is U.S. economy.

Table 9.2. The Correlation Coefficients of Temporary Shock in Technology

	Output	Consumption	Employment	Capital Stock
Model I Economy	0.9903	0.9722	0.9966	−0.0255
	(0.0031)	(0.0084)	(0.0013)	(0.1077)
Model IV Economy	0.8397	0.8510	0.4137	−0.1264
	(0.0512)	(0.0507)	(0.1862)	(0.1390)

Note: Numbers in parentheses are the corresponding standard deviations.

Next, we examine the temporary effect of the technology shock. Table 9.2 records the correlation coefficients of the temporary shock A_t from our 5,000 stochastic simulations. The two models predict rather different correlations. In the model I (RBC) economy, technology A_t has a temporary effect not only on consumption and output, but also on employment; all of them are strongly positive. Yet in our model IV economy with monopolistic competition and nonclearing labor market, we find that the correlation is much weaker with respect to employment. This is consistent with the widely discussed finding that technology has a near-0 (and even negative) effect on employment.

At the given expected market demand, an improvement in technology (reflected as an increase in labor productivity) will reduce the demand for labor if the firm follows the Keynesian way of determining output (i.e., the output is determined by demand). In this case, less labor is required to produce the given amount of output. Technical progress, therefore, may have an adverse effect on employment, at least in the short run.[8] This fact cannot be explained in the RBC framework because, at the given wage rate, the demand for labor is determined by the marginal product, which should be increased with the improvement in technology. This chapter thus demonstrates that if we follow the Keynesian way of determining quantity in a monopolistic competition model, the technology puzzle in standard market-clearing models will disappear.

9.4 Conclusions

In chapter 8, we showed how households may be constrained in buying consumption goods in the product market by the firm's actual demand for labor. Then a noncleared labor market was derived from a multiple-stage decision process of households in which we have neglected the fact that firms may also be demand constrained in the product market. The proposition in this chapter that shows the firms' constraint in the product market, explains this additional complication that can arise due to the interaction of the labor market and the product market constraints.

We showed in this chapter how the firms' constraints on the product market may explain the technology puzzle: that positive technology shocks may have only a weak effect on employment in the short run—a phenomenon inconsistent with equilibrium business cycle models, where technology shocks and employment are predicted to be positively correlated. This result was obtained in an economy with monopolistic competition, as in New Keynesian economics, where prices and wages are set by a monopolistic supplier and are sticky, resulting in an updating of prices and wages in which only a fraction of prices and wages are optimally set in each time period. We also introduced a nonclearing labor market, resulting from a multiple-stage decision problem, in which the households' constraint on the labor market spills over to the product market and the firms' constraint on the product market generates employment constraints. We showed that such a model better matches time series data of the U.S. economy.

9.5 Appendix: Proof of Proposition 7

Let $X_t = Z_t L_t$, with Z_t the permanent shock resulting purely from productivity growth, and L_t resulting from population growth. We assume that L_t has a constant growth rate μ, and hence Z_t follows the growth rate $(\gamma - \mu)$. The production function can be written as $Y_t = A_t Z_t^\alpha K_t^{1-\alpha} H_t^\alpha$, where H_t equals $N_t L_t$ and can be regarded as total labor hours.

Let us first consider the firm's willingness to supply Y_t^*, $Y_t^* = X_t y_t^*$, under the condition that the rental rate of capital r_t clears the capital market and the wage rate w_t is given. In this case, the firm's optimization problem can be expressed as

$$\max \ Y_t^* - r_t K_t^d - w_t H_t^d,$$

subject to

$$Y_t^* = A_t \left(Z_t\right)^\alpha \left(K_t^d\right)^{1-\alpha} \left(H_t^d\right)^\alpha .$$

The first-order condition tells us that

$$(1-\alpha) A_t \left(Z_t\right)^\alpha \left(K_t^d\right)^{-\alpha} \left(H_t^d\right)^\alpha = r_t \tag{9.28}$$

$$\alpha A_t \left(Z_t\right)^\alpha \left(K_t^d\right)^{1-\alpha} \left(H_t^d\right)^{\alpha-1} = w_t, \tag{9.29}$$

from which we can obtain

$$\frac{r_t}{w_t} = \left(\frac{1-\alpha}{\alpha}\right) \frac{H_t^d}{K_t^d}. \tag{9.30}$$

Since the rental rate of capital r_t is assumed to clear the capital market, we can thus replace K_t^d in the above equations with K_t. Since w_t is given, and therefore the demand for labor can be derived from (9.29),

$$H_t^d = \left(\frac{\alpha A_t}{w_t}\right)^{\frac{1}{1-\alpha}} (Z_t)^{\frac{\alpha}{1-\alpha}} K_t.$$

Dividing both sides of the above equation by X_t, and then reorganizing, we obtain

$$n_t^d = \frac{0.3}{\bar{N}} \left(\frac{\alpha A_t Z_t}{w_t}\right)^{\frac{1}{1-\alpha}} k_t.$$

We regard this labor demand as the demand when the firm's desired activities are carried out, which is the second equation in (9.26). Given this n_t^d, the firm's desire to supply y_t^* can be expressed as

$$\begin{aligned} y_t^* &= A_t k_t^{1-\alpha} (n_t^d \, \bar{N} / 0.3)^{\alpha} \\ &= A_t k_t \left(\frac{\alpha A_t Z_t}{w_t}\right)^{\frac{\alpha}{1-\alpha}}. \end{aligned} \tag{9.31}$$

This is equation (9.27) as expressed in the proposition.

Next, we consider the case that the firm's supply is constrained by the expected demand $\hat{Y}_t$, $\hat{Y}_t = X_t y_t$. In other words, $\hat{y}_t < y_t^*$ where y_t^* is given by (9.31). In this case, the firm's profit maximization problem is equivalent to the following minimization problem:

$$\min \; r_t K_t^d + w_t H_t^d,$$

subject to

$$\hat{Y}_t = A_t (Z_t)^{\alpha} \left(K_t^d\right)^{1-\alpha} \left(H_t^d\right)^{\alpha}. \tag{9.32}$$

The first-order condition will still allow us to obtain (9.30). Using equations (9.32) and (9.30), we obtain the demand for capital K_t^d and labor H_t^d as

$$K_t^d = \left(\frac{\hat{Y}_t}{A_t Z_t^{\alpha}}\right) \left[\left(\frac{w_t}{r_t}\right)\left(\frac{1-\alpha}{\alpha}\right)\right]^{\alpha}$$

$$H_t^d = \left(\frac{\hat{Y}_t}{A_t Z_t^{\alpha}}\right) \left[\left(\frac{w_t}{r_t}\right)\left(\frac{\alpha}{1-\alpha}\right)\right]^{1-\alpha}.$$

Dividing both sides of the two equations by X_t, we obtain

$$k_t^d = \left(\frac{\widehat{y}_t}{A_t}\right)\left[\left(\frac{w_t}{r_t Z_t}\right)\left(\frac{1-\alpha}{\alpha}\right)\right]^{\alpha} \tag{9.33}$$

$$n_t^d = \left(\frac{0.3\widehat{y}_t}{A_t \bar{N}}\right)\left[\left(\frac{r_t Z_t}{w_t}\right)\left(\frac{\alpha}{1-\alpha}\right)\right]^{1-\alpha}. \tag{9.34}$$

Since the real rental of capital r_t will clear the capital market, we can replace k_t^d in (9.33) with k_t. Substituting it into (9.34) for explaining r_t, we obtain

$$n_t^d = \left(\frac{0.3}{\bar{N}}\right)\left(\frac{\widehat{y}_t}{A_t}\right)^{1/\alpha}\left(\frac{1}{k_t}\right)^{(1-\alpha)/\alpha}.$$

This is the second equation in (9.26).

10

Reflections and Conclusions

In this book, we intend to contribute to the current research on stochastic dynamic macroeconomics. Stochastic dynamic modeling strategies have become important in macroeconomics. We argue that in order for such modeling strategies to explain real world economies more effectively, some Keynesian features should be introduced. We have shown that with such a perspective, macromodels can be enriched while it becomes possible to resolve important puzzles.

We have pursued dynamic macroeconomic model building by presuming dynamically optimizing agents, but we do not presume clearing of all markets at all times. To allow for nonclearing markets, we have introduced a multiple-stage adaptive optimization process in which agents can reoptimize when facing market constraints. We have argued that intertemporal behavior and dynamically optimizing agents do not contradict a Keynesian view on the working of modern macroeconomies.

Keynes in fact distinguishes between individual intertemporal optimal behavior and aggregate behavior—the latter, in his view, should be the focus of macropolicy. Concerning saving and consumption, he demonstrates that individual saving will be undertaken, among other motives, in order to build up assets for future consumption or to secure resources to carry out further investment.[1] On the other hand, in Keynes's view, aggregate consumption and saving have to strike a delicate balance between the past and the future. He argues that an increase in saving by a reduction of current consumption will reduce effective demand and employment, but it will have this effect only to the extent that it is not accompanied by an increase of expected future consumption and investment which stimulates current output and employment. Thus consumption is seen to represent a link between the past (past output and income) and the future.[2]

Investment is driven by prospective yields depending on future effective demand. Saving and investment lead to future consumption, and thus stimulation of investment and output, but they also lead to lower current consumption, thus endangering current demand and employment. Whether or not aggregate consumption reduction and increase in saving will be expansionary or contractionary depends on income expectations of households and on the expectations of the valuation of firms' capital assets. Concerning expectations, Keynes stresses that not only is the current state of capital equipment and technology due to various past states of expectations, but also that there may be multiple future long-run positions generated by the dynamics of expectations,[3] a topic that we pursued in chapter 7.

Moreover, although price and wage flexibility may be advantageous for the individual agent, aggregate behavior may be destabilized by the fluctuations of aggregate prices and wages. In fact, in Keynes, sticky wages and prices are perceived to be stabilizing. Although Keynes does not believe that large fluctuations in unemployment cannot be derived on the basis of the classical postulates (and are not voluntary or frictional), he is ambiguous on the causes of persistent unemployment—whether it comes from protracted periods of disequilibrium or is an equilibrium phenomenon (so that the economy is stuck at low-level equilibria). In response to such issues, in our modeling strategy of chapters 8 and 9 we have worked out model variants with sticky prices and wages, particularly with a exogenously given wage sequence, as perceived in Keynes. This helps us to improve on the labor market puzzles. Yet, as remarked above, an important future extension would be to make wage adjustments more endogenous, such as proposed in Uhlig (2004), in which wage rates respond to past wage rates and employment. In fact, a realistic modeling of wage and price setting is essential for future work on the labor market (see Gong and Semmler 2005), where wages are gradually adjusted toward optimal wages.

We also show that, based on our adaptive optimization behavior of economic agents, households' demand for consumption goods reflects optimal decisions concerning the future paths of technology and capital stock, but also responds to current employment; and firms are also likely to be constrained on the product market. The latter phenomenon was introduced in chapter 9, where, given demand constraints for firms on the product market, it appears reasonable that technology shocks have no or even negative effects on employment in the medium run. As we argue, this may help to solve the technology puzzle of standard RBC models in which technology shocks are procyclical and increase employment. In our context, although technology shocks are likely to be positively correlated with employment in the long run, in the short and medium runs they may have adverse effects on employment.

We know that presuming intertemporal optimizing behavior for aggregate behavior is not uncontroversial. It is, in particular, controversial when a single agent's dynamic optimization behavior is posited—as in the representative agent RBC model—also to hold for the aggregate behavior. As Keynes warns us, individual optimal choices do not necessarily result in socially preferred outcomes. Social aggregate choice may be needed to complement individual behavior.[4] Taking this into account, intertemporally optimizing behavior for households may be a useful assumption for aggregate behavior as well. As has been pointed out, this view can also provide an interesting framework[5] to study growth regimes, or eras, with certain preferences and technologies. As was discussed in chapter 8, for countries there may be distinct differences of time of growth regimes that suggest substantial changes of consumption, saving, and technology, the outcome of which may be measured by some welfare function in which more variables than the per capita income enter.

As mentioned in the introduction and referred to in chapters 8 and 9, research along the lines of Keynesian micro-oriented macroeconomics historically has been developed by two approaches: one is the tradition of a nonclearing market (or disequilibrium analysis), and the other is the New Keynesian analysis of monopolistic competition and sticky prices and wages. These two approaches are contrasted in chapters 8 and 9. We find that one can improve on the labor market and technology puzzles once we combine these two approaches.[6] We want to argue that the two traditions can indeed be complementary rather than mutually exclusive, and therefore they can be somewhat consolidated into a more complete system of price and quantity determinations within the Keynesian tradition. The main new method we are using here to reconcile the two traditions is a multiple-stage optimization behavior (adaptive optimization) in which agents reoptimize once they have perceived and learned about market constraints. Thus, adaptive optimization permits us to properly treat the market adjustment for nonclearing markets, which, we hope, allows us to make some progress to better match the model with time series data.

As for the quantitative and empirical dimensions of our study, we focus on approximate solutions that are computed from two types of first-order conditions: the Euler equation and the equations derived from a discrete-time Lagrangian approach. After evaluating other approximation methods, such as linearization or log-linearization of the first-order conditions, we propose a method that is developed from the work of Chow, for which the algorithm is written as a GAUSS procedure and is repeatedly used in the various chapters. Although recent methods sometimes suggest taking second-order approximations as well as evaluating policy shocks, in particular if the shocks are large (e.g., tax changes),[7] our method appears to be sufficiently accurate to obtain linear feedback rules so as to (a) calibrate the model variants, (b) pursue quantitative studies in terms of maximum likelihood and GMM estimations of the parameters, and (c) supply a multiple-stage decision model by which the working of labor and product markets can be understood when prices and wages are sticky.

Yet, occasionally we also use more advanced methods, such as dynamic programming with flexible grid size, as presented and applied in Grüne and Semmler (2004a, 2004b). In order to solve more general intertemporal deterministic and stochastic decision models, we have, in part I of the book, applied this new numerical algorithm, which is more appropriate for the solution of more general stochastic decision problems, such as model variants with a steep slope of the value function or a nondifferentiable value function in which decision variables may be discontinuous.

Finally, all our model variants, solved and studied by using either the Chow method or dynamic programming, are small-scale models. Yet, increasingly large-scale models are introduced in the macroeconomic literature—for example, for the study of monetary and fiscal policy effects—that are solved by first- and second-order approximations about a steady state. An important solution method of this kind is DYNARE, developed by Juillard (1996). Large-scale macromodels, in particular for policy evaluations, have recently made much progress in their application to the U.S. and European economies.[8]

Notes

Introduction and Overview

1. See also the volume in honor of Edmund Phelps, edited by Aghion et al. (2003), in which information problems are stressed. However, the concept of market clearing in recent New Keynesian literature is not unambiguous. We will discuss this in chapter 8.

2. Important papers in this line of research include Calvo (1983) and Rotemberg (1982). For recent reviews, see Taylor (1999) and Woodford (2003, chapter 3).

3. Keynes (1936) discusses the possibility of such an open-ended dynamics in chapter 5 of his book, in which he elaborates on the possibility of multiple long-period positions.

4. See Beyn et al. (2001) and Grüne and Semmler (2004a).

5. A recent evaluation of this failure of the RBC model is given in Schmidt-Grohe (2001).

6. Rotemberg and Woodford (1995, 1999), King and Wolman (1999), Gali (1999), and Woodford (2003) present models of monopolistic competition with price and wage stickiness.

7. For an evaluation of the search and matching theory, as well as of the role of shocks in explaining the evolution of unemployment in Europe, see Ljungqvist and Sargent (2003) and Blanchard (2003).

8. There is a long tradition of macroeconomic modeling with specification of nonclearing labor markets. See, for instance, Bénassy (1995, 2002), Malinvaud (1994), Danthine and Donaldson (1990, 1995), and Uhlig and Xu (1996). Although our approach owes a substantial debt to disequilibrium models, we move beyond this type of literature.

Chapter 1

1. Kendrick (1981) can be regarded as a seminal work in this field. For a review up to 1990, see Taylor and Uhlig (1990). For later developments, see Chow (1997), Judd (1998), Ljungqvist and Sargent (2000), and Marimon and Scott (1999). Recent methods of numerically solving the above discrete time Bellman equation (1.9) are presented in Santos and Vigo-Aguiar (1998) and Grüne and Semmler (2004a).

2. Named after Benveniste and Scheinkman (1979).

3. See also Chow (1997).

4. If we use (1.14) and (1.15) as the first-order condition, then there will be $2m + n$ equations, and y_t should include x_t, λ_t, and u_t.

5. Our suggested model here is a simplified version because we take only one lead. See also a similar formulation in Juillard (1996) with one lag and one lead.

6. The algorithm suggested by Chow (1993a, 1993b) is much more complicated. For any given x_t, the approximation as represented by (1.30)–(1.32) first occurs around (x_t, u^*), where u^* is the initial guess on u_t. Therefore, if u_t calculated via the decision rule (1.33), as the result of iterating (1.35)–(1.38), is different from u^*, then the approximation will occur again. However, this time it will be around (x_t, u_t), followed by iterating (1.35)–(1.38). The procedure will continue until the convergence of u_t. Since the above algorithm is designed for any given x_t, the resulting decision rule is nonlinear in x_t. In this sense, the method, as pointed out by Reiter (1997), is less convenient compared with other approximation methods. In a response to Reiter (1997), Kwan and Chow (1997) propose a one-time linearization around the steady states. Therefore, our above presentation follows the spirit of Kwan and Chow if we assume the linearization takes place around the steady states.

7. For instance, in GAUSS, one could use the procedure GRADP for deriving the partial derivatives.

8. The good approximation to $\frac{\partial F(x_t,u_t)}{\partial x_t}$ should be $\frac{\partial^2 F(\bar{x},\bar{u})}{\partial x^2}(x_t - \bar{x}) + \frac{\partial^2 F(\bar{x},\bar{u})}{\partial \bar{x}\partial u}(u_t - \bar{u}) + \frac{\partial F(\bar{x},\bar{u})}{\partial \bar{x}}$. For $\frac{\partial F(x_t,u_t)}{\partial u_t}$ there is a similar problem.

9. The same problem of multiple solutions should also exist for the log-linear approximation method.

10. However, multiple solutions may exist.

11. This is mostly the case in recent economic literature.

12. For a discretization of a continuous time, stochastic, optimal control problem with dynamics governed by an Itō stochastic differential equation, see Camilli and Falcone (1996).

13. A remark on this subject in an earlier version of our work was made in a personal communication by Buz Brock and Michael Woodford.

Chapter 2

1. See, for instance, Stokey et al. (1989, chapter 2), Blanchard and Fischer (1989, chapter 2), and Ljungqvist and Sargent (2000, chapter 2).

2. For a similar model where the depreciation rate is also set to 1, and hence the exact solution is computable, see Long and Plosser (1983).

3. The Benveniste–Scheinkman condition implies that the state variable does not appear in the transition equation. See chapter 1, section 1.3, of this book and Ljungquist and Sargent (2000, chapter 2).

4. The Euler equation is also essential not only in stochastic growth but also in finance, to study asset pricing, and in fiscal policy, to evaluate Treasury bonds and test for sustainability of fiscal policy. See Ljungqvist and Sargent (2000, chapters 2, 7, 10, and 17).

5. Reiter (1997) experienced the same problem.

6. Moreover, as concerns asset pricing, log-utility preferences provide us with a very simple stochastic discount factor and an analytical expression for the asset price. For $U(C) = \ln(C)$, the asset price is

$$P_t = E_t \sum_{t=1}^{\infty} \beta^j \frac{U'(C_{t+j})}{U'(C_t)}$$

$$= E_t \sum_{t=1}^{\infty} \beta^j \frac{C_t \cdot C_{t+j}}{C_{t+j}} = \frac{\beta C_t}{1-\beta}.$$

For further details, see Cochrane (2001, chapter 9.1) and Grüne and Semmler (2004b).

7. With 100 nodes in the capital stock interval, the error is $3.2 \cdot 10^{-2}$, and with 2,000 nodes the error shrinks to $6.3 \cdot 10^{-4}$. See Grüne and Semmler (2004a).

Chapter 3

1. See, e.g., Kydland and Prescott (1982), Long and Plosser (1983), Prescott (1986), Hansen (1985, 1988), King et al. (1988a, 1988b), and Plosser (1989). Recently, other statistics (in addition to the first and second moments) proposed by early business cycle literature—e.g., Burns and Mitchell (1946) and Adelman and Adelman (1959)—have been employed for this comparison. See King and Plosser (1994) and Simkins (1994).

2. For early critiques of the parameter selection employed in the calibration technique, see Singleton (1988) and Eichenbaum (1991).

3. See, e.g., Christiano and Eichenbaum (1992), Burnside et al. (1993), Chow (1993a, 1993b), and Chow and Kwan (1998).

4. The parameters in the state equation could be estimated independently.

5. For the conventional optimization algorithm, see appendix B of Judge et al. (1985) and Hamilton (1994, chapter 5).

6. This is from thermodynamics.

7. N_S is suggested to be 20 by Corana et al. (1987).

8. N_T is suggested to be 100 by Corona et. al. (1987).

9. R_T is suggested to be 0.85 by Corana et. al. (1987).

Chapter 4

1. See Arrow and Debreu (1954) and Debreu (1959).

2. We leave this exercise to the readers.

3. Indeed, they are essentially the same as the parameters chosen by King et al. (1988a, 1988b), except the last three parameters, which are related to the stochastic equation (4.25).

4. See Bennet and Farmer (2000) and Kim (2004).

5. See Benhabib and Nishimura (1998) and Harrison (2001).

6. See Kim (2003).

7. See Farmer (1999, chapter 7.2.4) and Benhabib and Farmer (1994).

8. For the following, see Lucas and Prescott (1971), Kim (2003), and Boldrin et al. (2001).

Chapter 5

1. The parameters in the exogenous equation could be independently estimated because they have no feedback into the other equations. Therefore, there is no necessity to include the exogenous equation in the restriction.

2. Here we use z_t rather than y_t in order to distinguish it from the output y_t in the model.

3. $\overline{N}$ can be regarded as the mean of per capita hours. γ does not appear in the model, but is created for transforming the model into a stationary version. The parameters with regard to the AR(1) process of A_t have no feedback effect on our estimation restrictions.

4. For a discussion of data definitions in RBC models, see Cooley and Prescott (1995).

5. We chose 0.0045 for γ to make k_t, c_t, and y_t stationary. (Note that we have defined $k_t \equiv \frac{K_t}{X_t}$, $c_t \equiv \frac{C_t}{X_t}$, and $y_t \equiv \frac{Y_t}{X_t}$.) Also, γ is the standard parameter chosen by King et al. (1988a).

6. These are the same as in table 4.1.

7. However, we were not able to obtain the same at the 1% significance level.

8. This is different from our previous notation N_t, which is the hours per capita.

9. A similar volatility is found in Burnside et al. (1996).

10. Here the structural parameters are still the standard ones given in table 5.4.

Chapter 6

1. See, for example, Jerman (1998), Boldrin et al. (2001), and Grüne and Semmler (2004b).

2. Using Christiano's data set, we implicitly assume that the standard model can, to some extent, replicate the moments of the real variables. Of course, as chapter 5 has shown, the standard model also fails along some real dimensions.

3. Using the thirty-day rate allows us to keep inflation uncertainty at a minimum.

4. See also Sharpe (1964).

5. As in our previous modeling, here we apply the power utility as describing the preferences of the representative household. For the model of asset market implications, other preferences (e.g., habit formation) are often

employed. See Jerman (1998), Boldrin et al. (2001), and Cochrane (2001, chapter 21), and Grüne and Semmler (2004b).

6. For further details, see Cochrane (2001, chapters 1.2 and 2.1). In RBC models the risk-free rate is generally too high, and its standard deviation is much too low, compared with the data. See Hornstein and Uhlig (2001).

7. Here we use the formula $Ee^x = e^{Ex+\sigma_x^2/2}$.

8. See Lettau and Uhlig (1999) for the details.

9. This value is advocated in Benninga and Protopapadakis (1990).

10. We constrain the estimates to lie between 0 and 1.

11. See also Wöhrmann et al. (2001), where time-varying characteristics of asset prices are explored.

12. See Grüne and Semmler (2004b,c).

Chapter 7

1. In Keynes (1936) such an open-ended dynamic is described in chapter 5. Keynes shows how higher or lower "long term positions" associated with higher or lower output and employment might be generated by expectational forces.

2. See, for example, Kim (2004).

3. Although these are important variants of macrodynamic models with optimizing behavior, it has recently been shown—see Beyn et al. (2001) and Grüne and Semmler (2004a)—that indeterminacy is likely to occur only at a point in these models (at a threshold), and not within a set, as the indeterminacy literature often claims.

4. Feichtinger et al. (2000) show that relative adjustment cost, where investment and capital stock enter the adjustment costs, is likely to generate multiple steady-state equilibria.

5. In the literature those thresholds have been called Skiba points (see Skiba 1978).

6. In particular, $k_t \equiv \frac{K_t}{X_t}$, $c_t \equiv \frac{C_t}{X_t}$, $i_t \equiv \frac{I_t}{X_t}$, $y_t \equiv \frac{Y_t}{X_t}$, and $q_t \equiv \frac{Q_t}{X_t}$.

Chapter 8

1. A number of important papers studying information issues relevant to modern macroeconomics are in the volume honoring Edmund Phelps's work (Aghion et al. 2003).

2. See Danthine and Donaldson (1990, 1995), Bénassy (1995), and Uhlig and Xu (1996), among others. A recently developed model of nonclearing markets of the French disequilibrium tradition, which resembles that of the United States, is in Portier and Puch (2004). Uhlig (2004) also presumes that models with exogenous wage sequences at the nonclearing market level will be better suited to match actual labor market movements.

3. The labor supply in these models is implicitly assumed to be given exogenously, and is normalized to 1.

4. Keynes (1936) not only had observed a widespread phenomenon of downward rigidity of wages but also had attributed strong stabilizing properties to wage stickiness.

5. One could perceive a change in secular forces concerning labor supply from the side of households—for example, changes in preferences, demographics, productivity and real wages, union bargaining, evolution of wealth, taxes, and subsidies—all of which affect labor supply. Some of those secular forces are often mentioned in the work by Phelps; see Phelps (1997) and Phelps and Zoega (1998). Recently, in Europe, generous unemployment compensation and related welfare state benefits have been added to the list of factors affecting the supply of labor, intensity of job search, and unemployment. For an extensive treatment of those factors, see Blanchard and Wolfers (2000) and Ljungqvist and Sargent (1998, 2003).

6. In Europe, on the demand side one could add, besides the pure technology shocks and the real wage, the roles of aggregate demand, high interest rates (Phelps 1997; Phelps and Zoega 1998), costs of hiring and firing, capital shortages, and slowdown of growth. See Malinvaud (1994) for a more extensive list of those factors.

7. Another line of recent research on modeling unemployment in a dynamic optimization framework can be found in Merz (1999), who employs search and matching theory to model the labor market; see also Ljungqvist and Sargent (1998, 2003). Yet, unemployment resulting from search and matching problems can be viewed as frictional unemployment (see Malinvaud 1994 for his classification of unemployment). As will become clear, this will be different from the unemployment that we discuss in this chapter. Also as shown, the search and matching models can capture neither the actual volatility of vacancies and unemployment nor the shifts in the unemployment rate (see Shimer 2003).

8. Keynes (1936, chapters 8–10) also stresses that consumption is linked to current output and income as well as to expected future output and income.

9. See, for example, Woodford (2003, chapter 3). There are also traditional Keynesian models that allow for disequilibria; see Bénassy (1984), among others. Yet, the well-known problem of these earlier disequilibrium models was that they disregard intertemporal optimizing behavior and never specify who sets the price. This has now been resolved by the modern literature of monopolistic competition, as in Woodford (2003). However, while resolving the price-setting problem, the decision with regard to quantities seems to be unresolved. The supplier may no longer behave optimally concerning its supply decision, but simply supply whatever quantity the market demands at the current price.

10. For models with multiple steps of optimization in the context of learning models, see Dawid and Day (2003), Sargent (1999), and Zhang and Semmler (2003).

11. For our simple representative agent model without money, this simplification does not affect our major result derived from our model. Meanwhile, it will allow us to save some effort to explain the nominal price determination, a focus in the New Keyensian literature.

12. See, for instance, Erceg et al. (2000), Christiano et al. (2001), and Woodford (2003).

13. These are basically the efficiency wage models mentioned in the introduction.

14. This is emphasized by Rotemberg (1982).

15. See the discussion in Christiano et al. (2001) and Zbaracki et al. (2000).

16. For such a treatment of wages as derivative security, see Uhlig (2003). For further details of the pricing of such liabilities, see Grüne and Semmler (2004c).

17. These are basically those prices which have not been adjusted for some periods, and their adjustment costs (such as the reputation cost) may not be high.

18. This type of wage setting is used in Woodford (2003, chapter 4) and Erceg et al. (2000).

19. For a reasonable calibration of a wage adjustment process generating a sticky labor market, see Uhlig (2004), and Gong and Semmler (2005).

20. Strictly speaking, the so-called labor market clearing should be defined as the condition that the firm's willingness to demand factors is equal to the household's willingness to supply factors. Such a concept has disappeared in the New Keynesian literature, in which the household supplies the labor effort according to the market demand and therefore does not seem to face excess demand or supply. Yet, even in this case, the household's willingness to supply labor effort is not necessarily equal to the actual supply or the market demand. At some point the marginal disutility of work may be higher than the preset wage. This indicates that even if there are no adjustment costs, so that the household can adjust the wage rate at every time period t, the disequilibrium in the labor market may still exist. In appendix I these points are illustrated in a static version of the working of the labor market.

21. This could also be realized by firms by demanding the same (or fewer) hours per worker but employing more workers than optimal. This case corresponds to what is discussed in the literature as labor hoarding—firms hesitate to fire workers during a recession because it may be hard to find new workers in the next upswing; see Burnside et al. (1993). In this case, firms may be off their marginal product curve, and thus this might require wage subsidies for firms, as has been suggested by Phelps (1997).

22. This could be achieved by employing the same number of workers but with each worker supplying more hours (varying shift length and overtime work); for a more formal treatment of this point, see Burnside et al. (1993).

23. Note that if firms are off their supply schedule and workers are off their demand schedule, a proper study would have to compute the firms' cost increase and profit loss, and the workers' welfare loss. If, however, the marginal cost for firms is rather flat (as empirical literature has argued; see Blanchard and Fischer 1989) and the marginal disutility is also rather flat, the overall loss may not be so high. The departure of the value function—as measuring the welfare of the representative household from the standard case—is studied in Gong and Semmler (2001). Results of this study are reported in appendix III of this chapter.

24. For a position representing this view, see Ljungqvist and Sargent (1998, 2003). For comments on this view, see Blanchard (2003); see also Walsh (2002), who employs search and matching theory to derive the persistence of real effects resulting from monetary policy shocks.

25. Hicks (1963) has called this frictional unemployment. Recently, one important form of mismatch in the labor market seems to be the mismatch of skills; see Greiner et al. (2004).

26. An alternative formulation for the household decision problem, as suggested by some authors, could simply employ the requirement that the marginal rate of substitution between consumption and labor effort is equated to the given real wage. In our perception this is equivalent to adding the constraint that $c_t^d = w_t n_t^s$. This way we then could derive the consumption demand and labor supply as functions not only of k_t and A_t but also of w_t. This would make the labor supply more consistent with the demand for labor, which also depends on the given wage rate, as shown above. Yet, this formulation may imply a difficulty in terms of the institutional background. In the RBC framework, the household is not only a worker but also a firm owner who makes investment decisions. The income of the household includes not only the wage but also profit. The restriction that consumption must equal wage income implies that the household is only a worker. The problems arising hereby may not be resolved satisfactorily if we do not introduce heterogeneous agents. See also chapter 8, section 8.4.4.

27. Note that X_t includes both population and productivity growth.

28. The negative response of labor effort to capital stock is also visible in the more general RBC model. See chapter 4, figure 4.4.

29. The property of consumption as representing the link between the past and the future can also be found in Keynes (1936, chapters 8–10).

30. However, here one might apply the efficiency wage theory or other theories, such as the staggered contract theory, to model the wage stickiness more accurately. An attempt to calibrate the actual real wage process for the United States is undertaken in Uhlig (2004). See also Gong and Semmler (2005).

31. This rescaling is necessary because we do not know the exact initial condition of Z_t, which we set equal to 1. We rescaled the wage series in such a way that the first observation of employment is equal to the demand for labor, as specified by equation (8.27).

32. Due to the discussion on Solow residual in chapter 5, we now see that A_t computed as the Solow residual may reflect the demand shock in addition to the technology shock.

33. See Uhlig (2004), where the implication for asset pricing is also spelled out.

34. In another study with German data, where we also introduce monopolistic competition in the product market (as in chapter 9), we find that ω is slightly positive.

35. We do not include the model III economy for calibration. Due to the value of the weighting parameter ω, the model III economy is equivalent to the model I economy. We thus report only results for the compromising rule (model I economy) and the short-side rule (model II economy).

36. See, for example, Nickell (1997), Nickell et al. (2003), and Meyers (1964).

37. See, for example, Uhlig and Xu (1996).

38. This is similar to Woodford's (2003, chapter 3) idea of a deviation of the efficient and natural level of output in which the efficient level is achieved only in a competitive economy with no frictions.

39. Ernst et al. (2004) also shows that the ω is strongly negatively correlated with variables representing rigidities of labor market institutions.

40. This could also be realized by firms by demanding the same (or fewer) hours per worker but employing more workers than optimal. The case would then correspond to what is discussed in the literature as labor hoarding, in which firms hesitate to fire workers during a recession because it may be hard to find new workers in the next upswing; see Burnside et al. (1993). In this case, firms may be off their marginal product curve, and this might require wage subsidies for firms, as has been suggested by Phelps (1997).

41. In fact, real wage rigidities in the United States are almost the same as in Europe. See Flaschel et al. (2001).

42. See Flaschel et al. (2001) and Smets and Wouters (2004).

43. See, for example, Nickell (1997), Nickell et al. (2003), and Meyers (1964).

44. It might reasonably be argued that, due to intertemporal optimization subject to the budget constraints, the supply specified by the decision rule may only approximate the decisions of those households for which unemployment is not expected to pose a problem for their budgets. Such households are more likely to be currently employed, represented by labor unions, and covered by legal restrictions. In other words, currently employed labor decides, through the optimal decision rule, about labor supply—and those who are currently unemployed do not. Such a feature could presumably be better studied by an intertemporal model with heterogeneous households. See, for example, Uhlig and Xu (1996).

45. This branch of literature is based on the seminal work by Merz (1995), Langot (1995), and Andolfatto (1996. More recent developments can be found in Cooley and Quadrini (1999), den Haan et al. (2000), and Ljungqvist and Sargent (1998, 2003).

46. For an evaluation of the search and matching theory, see Shimer (2003). For the role of labor market institutions and their interaction with shocks to explain the evolution of unemployment, see Blanchard and Wolfers (2000) and Blanchard (2003).

47. See Campbell (1994) for a modeling of a trend in technology shocks.

48. In the literature on long-term unemployment in Europe, using a CES production function, it has also been argued that there has been a long-run substitution of capital for labor, resulting from high real wages and high elasticity of substitution between capital and labor. See Caballero and Hammour (1997) and Hellwig (2004). Of course, the trend in the wage rate is also important in the equation for labor demand (equation (8.27)). For an account of the technology trend, see Flaschel et al. (2001), and for an additional account of the wage rate, see Heckman (2003).

49. Phelps and his coauthors have pointed out that an important change in the household's preferences in Europe is that households now rely also on assets instead of solely on labor income. For a model that stresses preferences and a lack of structural transformation in Europe toward the service sector, see Rogerson (2004).

50. Appendix III computes the welfare losses of our different model variants of a nonclearing labor market. There we find, similar to Ljungqvist

and Sargent (1998), but in contrast to Portier and Puch (2004), that the welfare losses are very small.

51. See Blanchard and Wolfers (2000), Greiner et al. (2004), and Heckman (2003). Yet it has been noted that although the per capita output in Europe lags behind the United States, the output per hour worked does not; see Blanchard (2004).

52. See also Portier and Puch (2004).

Chapter 9

1. For our simple representative agent model without money, this simplification does not affect our major result derived from our model. It will, however, allow us to avoid dealing with the nominal price determination, a main focus in the New Keynesian literature.

2. Here, when $y_t^s < \min(\widehat{y}_t, y_t^*)$, there will be insufficient inputs to produce $\min(\widehat{y}_t, y_t^*)$. On the other hand, when $y_t^s > \min(\widehat{y}_t, y_t^*)$, not all inputs will be used in production, and therefore resources are wasted.

3. Of course, one can also consider other forms of expectation. One possibility is to assume that the expectation is rational, so that it is equal to the steady state of y_t. Indeed, we have done the same empirical study with this assumption, but the result is less satisfying.

4. The algorithm used here is from chapter 1 of this volume.

5. There is a need for rescaling the wage series in the estimation of ω. This rescaling is necessary because we do not know the exact initial condition of Z_t, which we set equal to 1. We have followed the same rescaling procedure as in chapter 8.

6. Of course, for this exercise one should still consider A_t, the observed Solow residual, to include not only the technology shock but also the demand shock, among others.

7. This long-run effect of technology is also revealed by recent time series studies in the context of a variety of endogenous growth models. See Greiner et al. (2005).

8. A similar result was obtained by Engel and Matsumoto (2004), who also maintained that in a Keynesian framework, when the level of output is demand constrained, productivity shocks will have no effect on short-run output but the demand for labor will decline.

Chapter 10

1. See Keynes (1936), chapter 9.

2. Ibid., chapters 8 and 16.

3. Ibid., chapter 5.

4. Ibid., chapter 24.

5. An excellent framework for such a study is developed in Day and Young (2004).

6. Uhlig (2004) shows that one can also improve on the asset market puzzles if wage rates, employment, and consumption are made less endogenous.

7. See Schmidt-Grohe and Uribe (2004).

8. See, for example, Smets and Wouters (2004).

Bibliography

Adelman, I., and F. L. Adelman. (1959). "The Dynamic Properties of the Klein–Goldberger Model." *Econometrica* 27: 596–625.

Aghion, P., R. Frydman, J. Stiglitz, and M. Woodford, eds. (2003). *Knowledge, Information, and Expectations in Modern Macroeconomics*. Princeton, N.J.: Princeton University Press.

Andalfatto, D. (1996). "Business Cycles and Labor-Market Search." *American Economic Review* 86: 112–132.

Arrow, K. J., and G. Debreu. (1954). "Existence of an Equilibrium for a Competitive Economy." *Econometrica* 22: 265–290.

Asada, T., and W. Semmler. (1995). "Growth and Finance: An Intertemporal Model." *Journal of Macroeconomics* 17: 623–649.

Basu, S., J. Fernald, and M. S. Kimball. (1998). "Are Technology Improvements Contractionary?" Department of Economics, University of Michigan, Ann Arbor.

Basu, S., and M. S. Kimball. (1997). "Cyclical Productivity with Unobserved Input Variation." NBER Working Paper 5915. Cambridge, Mass.: National Bureau of Economic Research.

Bellman, R. (1957). *Dynamic Programming*. Princeton, N.J.: Princeton University Press.

Bénassy, J.-P. (1975). "Neo-Keynesian Disequilibrium Theory in a Monetary Economy." *Review of Economic Studies* 42: 502–523.

Bénassy, J.-P. (1982). *The Economics of Market Disequilibrium*. New York: Academic Press.

Bénassy, J.-P. (1995). "Money and Wage Contracts in an Optimizing Model of the Business Cycle." *Journal of Monetary Economics* 35: 303–315.

Bénassy, J.-P. (2002). *The Macroeconomics of Imperfect Competition and Nonclearing Markets.* Cambridge, Mass.: MIT Press.

Benhabib, J., and R. Farmer. (1994). "Indeterminacy and Increasing Returns." *Journal of Economic Theory* 63: 19–41.

Benhabib, J., and R. Farmer. (1999). "Indeterminacy and Sunspots in Macroeconomics." In *Handbook of Macroeconomics.* Vol. 1A, 387–448. Ed. J. Taylor and M. Woodford. New York: Elsevier.

Benhabib, J., and K. Nishimura. (1998). "Indeterminacy and Constant Returns." *Journal of Economic Theory* 81: 58–96.

Benhabib, J., S. Schmidt-Grohe, and M. Uribe. (2001). "Monetary Policy and Multiple Equilibria." *American Economic Review* 91, no. 1: 167–186.

Bennett, R. L., and R. E. A. Farmer. (2000). "Indeterminacy with Non-separable Utility." *Journal of Economic Theory* 93: 118–143.

Benninga, S., and A. Protopapadakis. (1990). "Leverage, Time Preference and the Equity Premium Puzzle." *Journal of Monetary Economics* 25: 49–58.

Benveniste, L. M., and J. A. Scheinkman. (1979). "On the Differentiability of the Value Function in Dynamic Economics." *Econometrica* 47, no. 3: 727–732.

Beyn, W. J., T. Pampel, and W. Semmler. (2001). "Dynamic Optimization and Skiba Sets in Economic Examples." *Optimal Control Applications and Methods* 22, no. 5–6: 251–280.

Blanchard, O. (2003). "Comments on Ljungqvist and Sargent." In *Knowledge, Information, and Expectations in Modern Macroceconomics,* 351–356. Ed. P. Aghion, R. Frydman, J. Stiglitz, and M. Woodford. Princeton, N.J.: Princeton Unviersity Press.

Blanchard, O. (2004). "The Economic Future of Europe." *Journal of Economic Perspectives* 18, no. 4: 3–26.

Blanchard, O., and S. Fischer. (1989). *Lectures on Macroeconomics.* Cambridge, Mass.: MIT Press.

Blanchard, O., and J. Wolfers. (2000). "The Role of Shocks and Institutions in the Rise of Unemployment: The Aggregate Evidence." *Economic Journal* 110: C1–C33.

Bohachevsky, I. O., M. E. Johnson, and M. L. Stein. (1986). "Generalized Simulated Annealing for Function Optimization." *Technometrics* 28: 209–217.

Bohn, H. (1995). "The Sustainability of Budget Deficits in a Stochastic Economy." *Journal of Money, Credit and Banking* 27, no. 1: 257–271.

Boldrin, M., L. Christiano, and J. Fisher. (1996). "Macroeconomic Lessons for Asset Pricing." NBER Working Paper 5262. Cambridge, Mass.: National Bureau of Economic Research.

Boldrin, M., L. Christiano, and J. Fisher. (2001). "Habit Persistence, Asset Returns and the Business Cycle." *American Economic Review* 91, no. 1: 149–166.

Brock, W. (1979). "An Integration of Stochastic Growth Theory and Theory of Finance, Part I: The Growth Model." In *General Equilibrium, Growth, and Trade,* 165–190. Ed. J. Green and J. Schenkman. New York: Academic Press.

Brock, W. (1982). "Asset Pricing in a Production Economy." In *The Economics of Information and Uncertainty,* 165–192. Ed. J. J. McCall. Chicago: University of Chicago Press.

Brock, W., and J. Mirman. (1972). "Optimal Economic Growth and Uncertainty: The Discounted Case." *Journal of Economic Theory* 4: 479–513.

Burns, A. F., and W. C. Mitchell. (1946). *Measuring Business Cycles.* New York: National Bureau of Economic Research.

Burnside, A. C., M. Eichenbaum, and S. T. Rebelo. (1993). "Labor Hoarding and the Business Cycle." *Journal of Political Economy* 101: 245–273.

Burnside, A. C., M. Eichenbaum, and S. T. Rebelo. (1996). "Sectoral Solow Residual." *European Economic Review* 40: 861–869.

Caballero, R. J., and H. L. Hammour. (1997). "Jobless Growth: Appropriability, Factor Substitution, and Unemployment." NBER Working Paper 6221. Cambridge, Mass.: National Bureau of Economic Research.

Calvo, G. A. (1983). "Staggered Contracts in a Utility Maximization Framework." *Journal of Monetary Economics* 12: 383–398.

Camilli, F., and M. Falcone. (1996). "Approximation of Optimal Control Problems with State Constraints: Estimates and Applications." In *Nonsmooth Analysis and Geometric Methods in Deterministic Optimal Control*, 23–57. IMA Volumes in Applied Mathematics 78. Ed. B. S. Mordukhovich and H. J. Sussmann. New York: Springer-Verlag.

Campbell, J. (1994). "Inspecting the Mechanism: An Analytical Approach to the Stochastic Growth Model." *Journal of Monetary Economics* 33: 463–506.

Capuzzo-Dolcetta, I. (1983). "On a Discrete Approximation of the Hamilton–Jacobi–Bellman Equation of Dynamic Programming." *Applied Mathematics and Optimization* 10: 367–377.

Chow, G. C. (1993a). "Statistical Estimation and Testing of a Real Business Cycle Model." Research Memorandum no. 365. Princeton, N.J.: Econometric Research Program, Princeton University.

Chow, G. C. (1993b). "Optimum Control without Solving the Bellman Equation." *Journal of Economic Dynamics and Control* 17: 621–630.

Chow, G. C. (1997). *Dynamic Economics: Optimization by the Lagrange Method.* New York: Oxford University Press.

Chow, G. C., and Y. K. Kwan. (1998). "How the Basic RBC Model Fails to Explain U.S. Time Series." *Journal of Monetary Economics* 41: 308–318.

Christiano, L. J. (1987). "Technical Appendix to 'Why Does Inventory Investment Fluctuate So Much?'" Research Department Working Paper 380. Minneapolis, Minn.: Federal Reserve Bank.

Christiano, L. J. (1988): "Why Does Inventory Fluctuate So Much?" *Journal of Monetary Economics* 21: 247–280.

Christiano, L. J., and M. Eichenbaum (1992). "Current Real Business Cycle Theories and Aggregate Labor Market Fluctuation." *American Economic Review* (June): 431–472.

Christiano, L. J., M. Eichenbaum, and C. Evans. (2001). "Nominal Rigidities and the Dynamic Effects of a Shock to Monetary Policy." Working Paper 107, Federal Reserve Bank of Cleveland.

Cochrane, J. H. (2001). *Asset Pricing.* Princeton, N.J.: Princeton University Press.

Cooley, T. F., and E. Prescott. (1995). "Economic Growth and Business Cycles." In *Frontiers of Business Cycle Research.* Ed. T. F. Cooley. Princeton, N.J.: Princeton University Press.

Cooley, T. F., and V. Quadrini. (1999). "A Neoclassical Model of the Phillips Curve Relation." *Journal of Monetary Economics* 44: 165–193.

Corana, A., M. C. Martini, and S. Ridella. (1987). "Minimizing Multimodal Functions of Continuous Variables with the Simulating Annealing Algorithm." *ACM Transactions on Mathematical Software* 13: 262–280.

Danthine, J. P., and J. B. Donaldson. (1990). "Efficiency Wages and the Business Cycle Puzzle." *European Economic Review* 34: 1275–1301.

Danthine, J. P., and J. B. Donaldson. (1995). "Non-Walrasian Economies." In *Frontiers of Business Cycle Research*. Ed. T. F. Cooley. Princeton, N.J.: Princeton University Press.

Dawid, H., and R. Day. (2003). "Sustainable Standard of Living and Adaptive Economizing in Economic Growth." Mimeo, Department of Economics, University of Bielefeld, Germany.

Day, R., and C. Young (2004). "Economic Growth and Revealed Social Preference." Mimeo, Department of Economics, University of Southern California, Los Angeles.

Debreu, G. (1959). *Theory of Value*. New York: Wiley.

den Haan, W., and A. Marcet. (1990). "Solving a Growth Model by Parameterizing Expectations." *Journal of Business Economics and Statistics* 8, no. 1: 31–34.

den Haan, W., G. Ramey, and J. Watson. (2000). "Job Destruction and Propagation of Shocks." *American Economic Review* 90, no. 3: 482–498.

Diebold, F., X. L. E. Ohanian, and J. Berkowitz. (1995). "Dynamic Equilibrium Economies: A Framework for Comparing Model and Data." Technical Working Paper 174. Cambridge, Mass.: National Bureau of Economic Research.

Eichenbaum, M. (1991). "Real Business Cycle Theory: Wisdom or Whimsy?" *Journal of Economic Dynamics and Control* 15: 607–626.

Eichenbaum, M., L. Hansen, and K. Singleton. (1988). "A Time Series Analysis of Representative Agent Models of Consumption and Leisure Under Uncertainty." *Quarterly Journal of Economics* 103(1): 51–78.

Eisner, R., and R. Stroz. (1963). "Determinants of Business Investment." In *Impacts of Monetary Policy*. Englewood Cliffs, N.J.: Prentice-Hall.

Engel, C., and A. Matsumoto. (2004). "Home Bias in Equities and New Open Economy Macroeconomics." Mimeo, Department of Economics, University of Wisconsin, Madison.

Erceg, C. J., D. W. Henderson, and A. T. Levin. (2000). "Optimal Monetary Policy with Staggered Wage and Price Contracts." *Journal of Monetary Economics* 46: 281–313.

Ernst, E., G. Gong, and W. Semmler. (2004). "Labor Market Institutions and Employment in Euro-area Countries." Mimeo, Department of Economics, Bielefeld University, Germany.

Evans, C. (1992). "Productivity Shock and Real Business Cycles." *Journal of Monetary Economics* 29: 191–208.

Fair, R. C. (1984). *Specification, Estimation, and Analysis of Macroeconometric Models*. Cambridge, Mass.: Harvard University Press.

Fair, R. C., and J. B. Taylor. (1983). "Solution and Maximum Likelihood Estimation of Dynamic Nonlinear Rational Expectation Models." *Econometrica* 21, no. 4: 1169–1185.

Falcone, M. (1987). "A Numerical Approach to the Infinite Horizon Problem of Deterministic Control Theory." *Applied Mathematics and Optimization* 15: 1–13.

Farmer, R. (1999). *Macroeconomics with Self-Fulfilling Expectations*. Cambridge, Mass.: MIT Press.

Feichtinger, G., F. H. Hartl, P. Kort, and F. Wirl. (2000). "The Dynamics of a Simple Relative Adjustment-Cost Framework." Mimeo, Department of Economics, University of Technology, Vienna, Austria.

Flaschel, P., G. Gong, and W. Semmler. (2001). "A Keynesian Macroeconometric Framework for Analyzing Monetary Policy Rules." *Journal of Economic Behavior and Organization* 46: 103–136.

Francis, N., and V. A. Ramey. (2001). "Is the Technology-Driven Real Business Cycle Hypothesis Dead? Shocks and Aggregate Fluctuations Revisited." Department of Economics, University of California, San Diego.

Francis, N., and V. A. Ramey. (2003). "The Source of Historical Economic Fluctuations: An Analysis Using Long-Run Restrictions." Department of Economics, University of California, San Diego.

Gali, J. (1999). "Technology, Employment, and the Business Cycle: Do Technology Shocks Explain Aggregate Fluctuation?" *American Economic Review* 89: 249–271.

Goffe, W. L., G. Ferrier, and J. Rogers. (1992). "Global Optimization of Statistical Function." In *Computational Economics and Econometrics*. Vol. 1. Ed. H. M. Amman, D. A. Belsley and L. F. Pau. Boston: Kluwer Academic.

Gong, G. (1997). "Three Essays in Macroeconomics." Ph.D. Diss., New School University, New York.

Gong, G., and W. Semmler. (2001). "Real Business Cycles with Disequilibirum in the Labor Market: A Comparison of the U.S. and German Economies." Working paper. Bielefeld, Germany: Center for Empirical Macroeconomics, Bielefeld University.

Gong, G., and W. Semmler. (2005). "Sticky Prices and Nonclearing Labor Market in Business Cycles." Mimeo, Department of Economics, New School University, New York.

Greiner, A., J. Rubart, and W. Semmler. (2004). "Economic Growth, Skill-biased Technical Change and Wage Inequality. A Model and Estimations for the U.S. and Europe." *Journal of Macroeconomics* 26: 597–627.

Greiner, A., W. Semmler, and G. Gong. (2005). *Forces of Economic Growth: A Time Series Perspective*. Princeton, N.J.: Princeton University Press.

Grüne, L. (1997). "An Adaptive Grid Scheme for the Discrete Hamilton–Jacobi–Bellman Equation." *Numer. Math.* 75: 1288–1314.

Grüne, L. (2003). "Error Estimation and Adaptive Discretization for the Discrete Stochastic Hamilton–Jacobi–Bellman Equation." Preprint, Department of Mathematics, University of Bayreuth, Germany. http://www.unibayreuth.de/departments/math/~lgruene/papers.

Grüne, L., and W. Semmler. (2004a). "Using Dynamic Programming for Solving Dynamic Models in Economics." *Journal of Economic Dynamics and Control* 28: 2427–2456.

Grüne, L., and W. Semmler. (2004b). "Solving Asset Pricing Models with Stochastic Dynamic Programming." Working paper. Bielefeld, Germany: Center for Empirical Macroeconomics, Bielefeld University.

Grüne, L., and W. Semmler. (2004c). "Asset Pricing—Constrained by Past Consumption Decisions." Working Paper, Department of Economics, New School University, New York.

Grüne, L., and W. Semmler. (2005). "Default Risk, Asset Pricing and Debt Control." *Journal of Financial Econometrics* 3, no. 1: 1–28.

Hall, R. E. (1988). "The Relation Between Price and Marginal Cost in U.S. Industry." *Journal of Political Economy* 96: 921–947.

Hamilton, J. D. (1994). *Time Series Analysis.* Princeton, N.J.: Princeton University Press.

Hansen, G. H. (1985). "Indivisible Labor and Business Cycles." *Journal of Monetary Economics* 16: 309–327.

Hansen, G. H. (1988). "Technical Progress and Aggregate Fluctuations." Working paper, Department of Economics, University of California, Los Angeles.

Hansen, L. P. (1982). "Large Sample Properties of Generalized Methods of Moments Estimators." *Econometrica* 50, no. 4: 1029–1054.

Hansen, L. P., and R. Jagannathan. (1991). "Restrictions on Intertemporal Marginal Rates of Substitution Implied by Asset Returns." *Journal of Political Economy* 99: 225–262.

Hansen, L. P., and K. J. Singleton. (1982). "Generalized Instrument Variables Estimation of Nonlinear Rational Expectations Models." *Econometrica* 50, no. 5: 1268–1286.

Harrison, S. G. (2001). "Indeterminacy with Sector-specific Externalities." *Journal of Economic Dynamics and Control* 25: 747–776.

Hayashi, F. (1982). "Tobin's Marginal q and Average q: A Neoclassical Interpretation." *Econometrica* 50: 213–224.

Heckman, J. (2003). "Flexibility and Creation: Job Lessons from the German Experience." In *Knowledge, Information, and Expectations in Modern Macroceconomics*, 357–393. Ed. P. Aghion, R. Frydman, J. Stiglitz, and M. Woodford. Princeton, N.J.: Princeton University Press.

Hellwig, M. (2004). "The Relation Between Real Wage Rates and Employment: An Intertemporal General Equilibrium Analysis." *German Economic Review* 5, no. 3: 236–297.

Hornstein, A., and H. Uhlig. (2001). "What Is the Real Story for Interest Rate Volatility?" *German Economic Review* 1, no. 1: 43–67.

Jerman, U. J. (1998). "Asset Pricing in Production Economies." *Journal of Monetary Economics* 41: 257–275.

Judd, K. L. (1996). "Approximation, Perturbation, and Projection Methods in Economic Analysis." In *Handbook of Computational Economics* 511–585. Ed. H. M. Amman, D. A. Kendrick, and J. Rust. New York: Elsevier.

Judd, K. L. (1998). *Numerical Methods in Economics.* Cambridge, Mass.: MIT Press.

Judge, G. G., W. E. Griffiths, R. C. Hill, and T. C. Lee. (1985). *The Theory and Practice of Econometrics.* 2nd ed. New York: Wiley.

Juillard M. (1996). "DYNARE: A Program for the Resolution and Simulation of Dynamic Models with Forward Variables Through the Use of a Relaxation Algorithm." Working Paper 9602. Paris: CEPREMAP.

Kendrick, D. (1981). *Stochastic Control for Economic Models.* New York: McGraw-Hill.

Keynes, J. M. (1936). *The General Theory of Employment, Interest and Money.* London: Macmillan.

Kim, J. (2003). "Indeterminacy and Investment and Adjustment Costs: An Analytical Result." Working paper. Department of Economics, University of Virginia, Charlottesville.

Kim, J. (2004). "Does Utility Curvature Matter for Indetermincy?" *Journal of Economic Behavior and Organization* 57(4): 421–429.

King, R. G., and C. I. Plosser. (1994). "Real Business Cycles and the Test of the Adelmans." *Journal of Monetary Economics* 33: 405–438.

King, R. G., C. I. Plosser, and S. T. Rebelo. (1988a). "Production, Growth and Business Cycles I: The Basic Neo-classical Model." *Journal of Monetary Economics* 21: 195–232.

King, R. G., C. I. Plosser, and S. T. Rebelo. (1988b). "Production, Growth and Business Cycles II: New Directions." *Journal of Monetary Economics* 21: 309–341.

King, R. G., and S. T. Rebelo. (1999). "Resuscitating Real Business Cycles." In *Handbook of Macroeconomics.* Vol. 1. Ed. J. B. Taylor and M. Woodford. New York: Elsevier Science.

King, R. G., and A. L. Wolman. (1999). "What Should the Monetary Authority Do When Prices Are Sticky?" In *Monetary Policy Rules.* Ed. J. Taylor. Chicago: University of Chicago Press.

Kwan, Y. K., and G. C. Chow. (1997). "Chow's Method of Optimum Control: A Numerical Solution." *Journal of Economic Dynamics and Control* 21: 739–752.

Kydland, F. E., and E. F. Prescott. (1982). "Time to Build and Aggregate Fluctuation." *Econometrica* 50: 1345–1370.

Langot, F. (1995). "Unemployment and Business Cycle: A General Equilibrium Matching Model." In *Advances in Business Cycle Research*, 287–326. Ed. P.-Y. Hénin. New York: Springer-Verlag.

Lettau, M. (1999). "Inspecting the Mechanism: The Determination of Asset Prices in the Real Business Cycle Model." Working Paper 1834. CEPR, Tilburg, Netherlands.

Lettau, M., G. Gong, and W. Semmler. (2001). "Statistical Estimation and Moment Evaluation of a Stochastic Growth Model with Asset Market Restriction." *Journal of Economic Behavior and Organization* 44: 85–103.

Lettau, M., and H. Uhlig. (1999). "Volatility Bounds and Preferences: An Analytical Approach." Revised from Discussion Paper 1678. CEPR, Tilburg, Netherlands.

Ljungqvist, L., and T. Sargent. (1998). "The European Unemployment Dilemma." *Journal of Political Economy* 106, no. 3: 514–550.

Ljungqvist, L., and T. J. Sargent. (2000). *Recursive Macroeconomics Theory.* Cambridge, MA: MIT Press.

Ljungqvist, L., and T. Sargent. (2003). "European Unemployment: From a Worker's Perspective." In *Knowledge, Information, and Expectations in Modern Macroceconomics*, 326–350. Ed. P. Aghion, R. Frydman, J. Stiglitz, and M. Woodford. Princeton, N.J.: Princeton University Press.

Long, J. B., and C. I. Plosser. (1983). "Real Business Cycles." *Journal of Political Economy* 91: 39–69.

Lucas, R. (1976). "Econometric Policy Evaluation: A Critique." *Carnegie-Rochester Conference Series on Public Policy* 1: 19–46.

Lucas, R. (1978). "Asset Prices in an Exchange Economy." *Econometrica* 46: 1429–1446.

Lucas, R., and E. C. Prescott. (1971). "Investment Under Uncertainty." *Econometrica* 39, no. 5: 659ff.

Lucas, R. E. (1967). "Adjustment Costs and the Theory of Supply." *Journal of Political Economy* 75: 321–334.

Malinvaud, E. (1994). *Diagnosing Unemployment.* New York: Cambridge University Press.

Mankiw, N. G. (1989). "Real Business Cycles: A New Keynesian Perspective." *Journal of Economic Perspectives* 3: 79–90.

Marimon, R., and A. Scott. (1999). *Computational Methods for the Study of Dynamic Economies.* New York: Oxford University Press.

Mehra, R., and E. C. Prescott. (1985). "The Equity Premium Puzzle." *Journal of Monetary Economics* 15: 145–161.

Merz, M. (1995). "Search in the Labor Market and the Real Business Cycle." *Journal of Monetary Economics* 36: 269–300.

Merz, M. (1999). "Heterogeneous Job-Matches and the Cyclical Behavior of Labor Turnover." *Journal of Monetary Economics* 43: 91–124.

Metropolis, N., A. W. Rosenbluth, M. N. Rosenbluth, A. M. Teller, and E. Teller. (1953). "Equation of State Calculation by Fast Computing Machines." *Journal of Chemical Physics* 21, no. 6: 1087–1092.

Meyers, R. J. (1964). "What Can We Learn from European Experience?" In *Unemployment and the American Economy?* Ed. A. M. Ross. New York: Wiley.

Newey, W. K., and K. D. West. (1987). "A Simple, Positive Semi-definite, Heteroskedasticity and Autocorrelation Consistent Covariance Matrix." *Econometrica* 55: 195–232.

Nickell, S. (1997). "Unemployment and Labor Market Rigidities—Europe versus North America." *Journal of Economic Perspectives* 3: 55–74.

Nickell, S., L. Nunziata, W. Ochel, and G. Quintini. (2003). "The Beveridge Curve, Unemployment, and Wages in the OECD from the 1960s to the 1990s." In *Knowledge, Information, and Expectations in Modern Macroeconomics.* Ed. P. Aghion, R. Frydman, J. Stiglitz, and M. Woodford. Princeton, N.J.: Princeton University Press.

OECD. (1998a). "Business Sector Data Base." In OECD-Statistical Compendium, Paris, France.

OECD. (1998b). "General Economic Problems." In *OECD Economic Outlook.* OECD, Paris, France.

Phelps, E. (1997). *Rewarding Work.* Cambridge, Mass.: Harvard University Press.

Phelps, E., and G. Zoega. (1998). "Natural Rate Theory and OECD Unemployment." *Economic Journal* 108 (May): 782–801.

Plosser, C. I. (1989). "Understanding Real Business Cycles." *Journal of Economic Perspectives* 3, no. 3: 51–77.

Portier, F., and L. A. Puch. (2004). "The Welfare Cost of Business Cycles in an Economy with Nonclearing Markets." Working Paper, Department of Economics, University of Toulouse, France.

Prescott, E. (2003). "Why Do Americans Work So Much More Than the Europeans?" Research Department Staff Report 321. Minneapolis, Minn.: Federal Reserve Bank of Minneapolis.

Prescott, E. C. (1986). "Theory Ahead of Business Cycle Measurement." *Quarterly Review, Federal Reserve Bank of Minneapolis* 10, no. 4: 9–22.

Ramsey, F. P. (1928). "A Mathematical Theory of Saving." *Economic Journal* 28: 543–559.

Reiter, M. (1997). "Chow's Method of Optimum Control." *Journal of Economic Dynamics and Control* 21: 723–737.

Rogerson, R. (2004). "Structural Transformation and the Deterioration of European Labor Market Outcomes." Working Paper, Department of Economics, Arizona State University, Tempe.

Rotemberg, J. (1982). "Sticky Prices in the United States." *Journal of Political Economy* 90: 1187–1211.

Rotemberg, J., and M. Woodford. (1995). "Dynamic General Equilibrium Models with Imperfectly Competitive Product Markets." In *Frontiers of Business Cycle Research*. Ed. T. F. Cooley. Princeton, N.J.: Princeton University Press.

Rotemberg, J., and M. Woodford. (1999). "Interest Rate Rules in an Estimated Sticky Price Model." In *Monetary Policy Rules*. Ed. J. Taylor. Chicago: University of Chicago Press.

Rust, J. (1996). "Numerical Dynamic Programming in Economics." In *Handbook of Computational Economics*, 620–729. Ed. H. M. Amman, D. A. Kendrick, and J. Rust. New York: Elsevier.

Santos, M. S., and J. Vigo-Aguiar. (1998). "Analysis of a Numerical Dynamic Programming Algorithm Applied to Economic Models." *Econometrica* 66, no. 2: 409–426.

Sargent, T. (1987). *Dynamic Macroeconomic Theory*. Cambridge, Mass.: Harvard University Press.

Sargent, T. (1999). *Conquest of American Inflation*. Princeton, N.J.: Princeton University Press.

Schmidt-Grohe, S. (2001). "Endogenous Business Cycles and the Dynamics of Output, Hours and Consumption." *American Economic Review* 90, no. 5: 1136–1159.

Schmidt-Grohe, S., and H. Uribe. (2004). "Solving Dynamic General Equilibrium Models Using a Second-Order Approximation to the Policy Function." *Journal of Economic Dynamics and Control* 28: 755–775.

Sharpe, W. (1964). "Capital Asset Prices: A Theory of Market Equilibrium Under Conditions of Risk." *Journal of Finance* 19: 425–442.

Shimer, R. (2003). "The Cyclical Behavior of Equilibrium Unemployment and Vacancies: Evidence and Theory." NBER working paper 9536. Cambridge, Mass.: National Bureau of Economic Research.

Simkins, S. P. (1994). "Do Real Business Cycle Models Really Exhibit Business Cycle Behavior?" *Journal of Monetary Economics* 33: 381–404.

Singleton, K. (1988). "Econometric Issues in the Analysis of Equilibrium Business Cycle Model." *Journal of Monetary Economics* 21, 361–386.

Skiba, A. K. (1978). "Optimal Growth with a Convex-Concave Production Function." *Econometrica* 46 (May): 527–539.

Smets, F., and R. Wouters. (2004). "Comparing Shocks and Frictions in U.S.

and Euro Area Business Cycles. A Bayesian DSGE Approach." Working paper 391. European Central Bank, Frankfurt, Germany.

Solow, R. (1979). "Another Possible Source of Wage Stickiness." *Journal of Macroeconomics* 1: 79–82.

Statistisches Bundesamt. (1998). *Fachserie* 18. Wiesbaden, Germany: Statistisches Bundesamt.

Stokey, N. L., R. E. Lucas, and E. C. Prescott. (1989). *Recursive Methods in Economics.* Cambridge, Mass.: Harvard University Press.

Summers, L. H. (1986). "Some Skeptical Observations on Real Business Cycles Theory." *Federal Reserve Bank of Minneapolis Quarterly Review* 10: 23–27.

Taylor, J. B. (1980). "Aggregate Dynamics and Staggered Contracts." *Journal of Political Economy* 88: 1–24.

Taylor, J. B. (1999). "Staggered Price and Wage Setting in Macroeconomics." In *Handbook of Macroeconomics* Vol. 1. Ed. J. B. Tayor and M. Woodford. New York: Elsevier Science.

Taylor, J. B., and Uhlig, H. (1990). "Solving Nonlinear Stochastic Growth Models: A Comparison of Alternative Solution Methods." *Journal of Business and Economic Statistics* 8: 1–17.

Uhlig, H. (1999). "A Toolkit for Analysing Nonlinear Dynamic Stochastic Models Easily." In *Computational Methods for the Study of Dynamic Economies.* Ed. R. Marimon and A. Scott. New York: Oxford University Press.

Uhlig, H. (2003). "Do Technology Shocks Lead to a Fall of Total Hours Worked?" *Journal of the European Economic Association* 2, no. 2–3: 361–371.

Uhlig, H. (2004). "Macroeconomics and Asset Markets: Some Mutual Implications." Working Paper, Department of Economics, Humboldt University, Berlin.

Uhlig, H., and Y. Xu. (1996). "Effort and the Cycle: Cyclical Implications of Efficiency Wages." Mimeo, Department of Economics, Tilburg University, Netherlands.

Uzawa, H. (1968). "The Penrose Effect and Optimum Growth." *Economic Studies Quarterly* 19: 1–14.

Vanderbilt, D., and S. G. Louie. (1984). "A Monte Carlo Simulated Annealing Approach to Optimization over Continuous Variables." *Journal of Computational Physics* 56: 259–271.

Walsh, L. E. (2002). "Labor Market Search and Monetary Shocks." Working Paper, Department of Economics, University of California, Santa Cruz.

Watson, M. W. (1993). "Measures of Fit for Calibration Models." *Journal of Political Economy* 101, no. 6: 1011–1041.

Wöhrmann, P., W. Semmler, and M. Lettau. (2001). "Nonparametric Estimation of Time-Varying Characteristics of Intertemporal Asset Pricing Models." Working paper. Bielefeld, Germany: Center for Empirical Macroeconomics, Bielefeld University.

Woodford, M. (2003). *Interest and Prices.* Princeton, N.J.: Princeton University Press.

Zbaracki, M. J., M. Ritson, D. Levy, S. Dutta, and M. Bergen. (2000). "The Managerial and Customer Costs of Price Adjustment: Direct Evidence from Industrial Markets." Manuscript, Department of Economics, Wharton School, University of Pennsylvania, Philadelphia.

Zhang, W., and W. Semmler. (in press). "Monetary Policy Rules Under Uncertainty: Adaptive Learning and Robust Control." *Macroeconomic Dynamics*.

Index